Faulkner's Search for a South

WALTER TAYLOR

Faulkner's Search for a South

UNIVERSITY OF ILLINOIS PRESS
Urbana Chicago London

Manufactured in the United States of America

Library of Congress Cataloging in Publication Data

Taylor, Walter, 1927–
Faulkner's search for a South.

Bibliography: p.
Includes index.
1. Faulkner, William, 1897–1962—Criticism and
interpretation. 2. Southern States in literature.
I. Title.
PS3511.A86Z975 813'.52 81-16127
ISBN 0-252-00943-6 AACR2

TO
My wife
NANCY SEWARD TAYLOR

Contents

Preface

Faulkner set out in *Sartoris*, he said, "to recreate between the covers of a book the world." That world was the Mississippi of his youth, a world that at age twenty-nine he had found he "was already preparing to lose and regret"; his motive for writing was "the capture of that world and the feeling of it."[1]

By the 1950s Faulkner had established himself as the South's most widely respected and articulate literary voice, but he had also begun to speak of his career in a fashion that startled many readers. Asked to rate himself and his contemporaries, he responded that he could only do so in terms of "the magnificence, the splendor of the failure." The search for "perfection" was a lifelong, and fated, quest for a writer. "I think that the writer must want primarily perfection, that that is his one chance while he has breath, to attain perfection." But no writer could achieve it. "He can't do it in his life because he can't be as brave as he wishes he might, he can't always be as honest as he wishes he might." That was why a writer kept writing. With each new effort the writer could tell himself that "here's the chance to hope that when he has pencil and paper he can make something as perfect as he dreamed it to be." There was a relationship between a writer's failure with one work and his need to write the next. "If he wrote one book and that was all he had hoped of it, he would probably quit. But it's not, so he tries again and so he begins to think of his own work as a long series of failures." Perfection "is what he wants, and anything less than perfection is failure."[2]

Faulkner's failures would be triumphs for most writers; but his notions about them constitute more than a confessional. In these statements he has provided clues to a dynamic that helps to explain his career.[3] Gary Lee Stonum, drawing on Geoffrey Hartman,[4] has formulated an approach suitable for a career study of Faulkner. "The career [of a writer] does not proceed as a linear development or a gradual evolution of its own latent tendencies," Stonum writes. "Rather it develops by expressly questioning the assumptions on which the earlier work depends." This description seems especially apt for Faulkner, whose "writings," Stonum points out, "manifest a particular kind of intertextuality among themselves. . . . The exploitation of one set of assumptions within a given text raises up the challenge to these assumptions which becomes the basis of the next moment in the career. . . . This concatenation of texts . . . according to how they question their predecessors and make way for their successors is precisely the trajectory of Faulkner's career and the principal result of the kind of discipline to which it adheres."[5] A career, in short, can be a work of art in somewhat the same sense that a novel or a poem is a work of art, and that, in some fashion, is how Faulkner thought of his: as a career which "adheres" to a "discipline" that results in a "trajectory"—a career, in short, that has its own dynamic.

Stonum, although commenting that "some of the phases of Faulkner's career can be roughly defined by changes in the pressure exerted on him by tradition, locality, and also language,"[6] does not probe the significance of Faulkner's *southernness* for his career. I argue here that the image of "perfection" Faulkner was seeking was usually—perhaps always—something that could be described as "the South": some cluster of images, experiences, and fantasies inherited from the world of his youth. I suggest that each novel about "the South" is an effort to achieve "the capture of that world and the feeling of it," arguing that Faulkner's protagonists—Bayard Sartoris, Quentin Compson, and Isaac McCaslin—are not so much surrogates for Faulkner as they are representatives of possible responses to that world, responses from which Faulkner's art is an effort to exorcise himself, failures from which he frees himself in order to move on to the next vision.

Faulkner's career thus reveals an inner dynamic in which each work may be seen as part of a progressive effort to imagine what "the South" might have been, or might become, in both its benign and nightmarish aspects—and to imagine a series of protagonists who cope, or fail to cope, with it. This dynamic identifies a lifelong psychodrama that had its beginnings before Faulkner's birth and, to the degree that we are still struggling to understand it, has yet to end. It offers certain answers to the questions raised by the phenomenon that we describe as Faulkner's career and hence to the riddle of Faulkner's southernness. It is to that riddle that the present study is addressed.

Acknowledgments

I am indebted to many people for their help with this book, and it is impossible to acknowledge all of them. In particular, I should like to express my gratitude to the following: To the staff of the Library of the University of Texas at El Paso; to Florence Dick, a typist who was also a most helpful critic; to Floyd C. Watkins and Albert E. Stone, Jr., under whom I began my study of Faulkner some years ago; to the chairmen of the Department of English at the University of Texas at El Paso, John O. West, Tony J. Stafford, and James K. P. Mortensen, for encouragement and assistance; to Lawrence J. Johnson for invaluable help in reading portions of the manuscript; to my wife, Nancy Seward Taylor, for encouragement, for criticism, for close reading of manuscript and proofs, and for the many other things she did without which this work would not have been possible; to my mother, Florence Ferguson Taylor, for faith and for bibliographical and critical assistance; and to my father, Walter Fuller Taylor, Sr., who taught me scholarship by precept and example.

Abbreviations of Works by Faulkner

AA	*Absalom, Absalom!* (New York: Random House, 1936).
AILD	*As I Lay Dying* (New York: Random House, 1930).
CS	*Collected Stories of William Faulkner* (New York: Random House, 1950).
EPP	*William Faulkner: Early Prose and Poetry,* ed. Carvel Collins (Boston: Little Brown, and Co., 1962).
FID	*Flags in the Dust* (New York: Random House, 1973).
GDM	*Go Down, Moses* (New York: Random House, 1942).
H	*The Hamlet* (New York: Random House, 1940).
ID	*Intruder in the Dust* (New York: Random House, 1948).
LIA	*Light in August* (New York: Random House, 1932).
MFGB	*The Marble Faun and a Green Bough* (New York: Random House, 1929).
MOS	*Mosquitoes* (New York: Liveright, 1927).
NOS	*New Orleans Sketches* (Tokyo: Holuseido, 1953).
RN	*Requiem for a Nun* (New York: Random House, 1951).
S	*Sanctuary* (New York: Random House, 1931).
Sar	*Sartoris* (New York: Harcourt, Brace, 1929).
SF	*The Sound and the Fury* (New York: Random House, 1929).
SP	*Soldiers' Pay* (New York: Liveright, 1926).
TM	*The Mansion* (New York: Random House, 1959).
TR	*The Reivers* (New York: Random House, 1962).
U	*The Unvanquished* (New York: Random House, 1938).

1

A Nice Place to Live

In *A Green Bow* (1933) William Faulkner imagined his return to the Mississippi earth. That would be a kind of immortality, he wrote,

> . . . for where is any death
> While in these blue hills slumbrous overhead
> I'm rooted like a tree? Though I be dead,
> This earth that holds me fast will find me breath (*MFGB*, 67).

Mississippi lawyer Phil Stone had thought much about his friend's roots. "It is impossible to understand a tree," he wrote in 1934, "without knowing its source, the soil in which it grows, and the nature of the country which surrounds it." Understanding Faulkner meant understanding Mississippi. For Stone, it was important that "in the blood of William Faulkner . . . runs, whether he will or no, the blood of those who came before him"; that meant that in Faulkner's "thoughts, his emotions, the very gestures of his body, whether he will or no, move millions of unconscious reflections of the land which gave him birth." Stone was not surprised that Mississippi had fostered greatness; it was a land without extremes, "a country of not a great deal of riches and of almost no poverty," and living there had allowed Faulkner a pastoral, almost a Wordsworthian, youth. "From any part of Oxford," Stone wrote, "it is only a little walk to numerous places where one can find the unspoiled golden peace of legendary days and where the sound of mankind's so-called progress comes only dreamlike and from afar." In Oxford Faulkner was "a simple-hearted country boy leading the life of a country squire."[1]

Other southerners were less enthusiastic about their life-style. Two years before Faulkner was born, Booker T. Washington addressed the Cotton States Industrial Exposition in Atlanta. "The wisest among my race," he said, "understand that the agitation of questions of social equality is the extremest folly"; it was clear now that black "progress" would never come through "artificial forcing." Blacks ought to have "all privileges of the law"; but to Washington in 1895, other things seemed more pressing. Blacks had to "be prepared for the exercise of these privileges," he said, and they had to think about survival. "The opportunity to earn a dollar in a factory just now is worth infinitely more than the opportunity to spend a dollar in an opera-house."[2]

Not every southern black was ready to agree, but in one respect Washington was speaking for all of them; his Atlanta Exposition Address was an acknowledgment from the South's most widely accepted black leader of the end of the bright hopes fostered forty years earlier by Reconstruction. The trouble was that Washington seemed to be conceding more than that. Reconstruction had ended when federal troops withdrew after the Hayes-Tilden compromise of 1877, but the southern aristocrats who had engineered the compromise never dreamed that they could remake the South after its prewar mold. They had struggled back to power by intimidating blacks, carpetbaggers, and anyone else who stood in their way, but the next year blacks were still voting, still holding office, and still taking their grievances to court. And now here was Washington, less than twenty years later, acknowledging the advent of Jim Crow segregation.

Something had happened in Stone's land without extremes, and Washington's address signaled it. By 1911, when Faulkner was more than halfway through that pastoral boyhood Stone boasted of, Episcopal minister Edgar Gardner Murphy attempted to identify it. Murphy saw a "new mood" in the South, and he found it appalling. "Its spirit" was "that of an all-absorbing autocracy of race." It was driven by "an animus of aggrandizement which makes, in the imagination of the white man, an absolute identification of the stronger race with the very being of the state."[3] A new era had dawned. It had arrived just before Faulkner was born, and its heyday was precisely the span of that Wordsworthian youth that had allowed Stone and Faulkner to enjoy the "peace of legendary days" in a place "where the sound of mankind's so-called progress comes only dreamlike and from afar."

"Perhaps the most insidious . . . form of segregation," Ralph Ellison wrote years later, was "that of the word." There was "an inescapable connection between the writer and the beliefs and attitudes current in his culture," a connection that could be fatal for his art. To get his "inspiration," Ellison believed that every artist "must perpetually descend" into his unconscious to deal with "all the unconquered anguish of his living" that he found there. For white writers, Ellison thought, that could be tragic. Whites had created "the Negro stereotype" to avoid such probing. It was "an image of the unorganized, irrational forces in American life," and, when the white writer descended into his unconscious "to re-encounter . . . the ghosts of his former selves," it surfaced, smothering his perceptions "with the anesthesia of legend, myth, . . . and narcotic modes of thinking." That was the "segregation . . . of the word." The word had the power of life, but it also had a "subtle power to suggest . . . overt action while magically disguising the moral consequences of that action." Ellison believed that since the Hayes-Tilden compromise "the race issue has been like a stave driven into the American system of values."[4]

For the youthful Faulkner, no less than for Stone, it was "impossible to understand a tree without knowing its source"; understanding himself was going to mean understanding Mississippi. But what was Mississippi? Was it the pastoral haven Stone spoke of, where "a simple-hearted country boy" could lead "the life of a country squire"? Or was it the "all-absorbing autocracy of race" that Murphy foresaw? There was more than one South, and the South one found was not always the South one expected.

2

The Segregation of the Word

I

Black leader W. E. B. Du Bois was aware that there were many Souths, and he counseled caution. "The attitude of the Southern whites," he warned in 1903, "is not . . . in all cases the same." Du Bois saw "forces of all kinds . . . fighting for supremacy." Some were economic. There were "money makers" who were applying pressure "especially in country districts," where the black was now threatened with "semi-slavery"; there were "the workingmen" who "fear his competition." Whatever the economic factors, Harvard-trained Du Bois was sure that the common denominator was ignorance. The source of the problem was "the ignorant Southerner" who "hates the Negro." It was the ignorant masses who had allied with "those of the educated who fear the Negro" in an effort "to disfranchise him"; and it was "the passions of the ignorant" that were "easily aroused to lynch and abuse any black man." But Du Bois knew other whites who were neither ignorant nor fearful, "who wish to help [the Negro] to rise." These were "usually the sons of the masters," and it was "this last class" that, with the help of "National opinion," had managed "to maintain the Negro common schools, and to protect the Negro partially in property, life, and limb." Blacks ought not to forget that, he cautioned. It was "the duty of black men to judge . . . discriminatingly" and to remember that "the South is not 'solid.' "[1]

Du Bois's words held a special irony for Mississippians. Since Reconstruction the two white factions he identified had been locked in a bitter struggle. The "educated" — businessmen-aristocrats who were "sons of the masters" — were based in the Delta, a fertile lowland between Memphis and the mouth of the Yazoo at Vicksburg; the "ignorant" — emerging redneck farmers — were based in the less fertile northeastern hills. It was a classic American confrontation between Cavalier and Puritan. According to social myth the Delta's vast antebellum plantations were built by the sons of Virginia, Carolina, and Louisiana aristocrats; Deltans were elitist, high-church, and conservative. Hill men were poor-white farmers whose fathers, legend attests, had crossed the Appalachians as pioneers; they were fundamentalist in religion and agrarian in politics, militantly antirailroad and anticorporation.[2] Mississippi's two dominant white factions thus stood for inimical life-styles, and their struggle was a struggle over the nature of southern reality; it was a struggle re-enacted at each election for desperate, escalating stakes, and, as it intensified, it focused on those other Mississippians Du Bois was addressing.

The one issue that white Mississippians did agree on was Reconstruction. As an electorate, Mississippi's freedmen had performed as responsibly as most electorates of the era,[3] but whites remembered only the brief period of anarchy after the surrender. Law and order, they were convinced, had not returned until after the "Redemption": the Hayes-Tilden compromise of 1877 that resulted in the withdrawal of federal troops. Their hero was Democratic Senator Lucius Quintus Cincinnatus Lamar of Oxford, Mississippi, who aspired to the aristocratic image of the generation preceding him; he had "never made popularity the standard of . . . [his] action," Lamar insisted.[4] But hill farmers outnumbered aristocrats, and Lamar and his "Redeemers" were vulnerable to any independent ticket that could appeal to both blacks and hill men. In the Delta, where blacks outnumbered whites, aristocrats manipulated black voters by badgering and intimidation, but it was soon evident that to win white voters they would have to fall back on the one issue that could unite the party—the fear of black anarchy. "The safety of Mississippi," Lamar pronounced, "lies in the maintenance of the Democratic organization."[5] That election of 1875 set the tone for the three decades that followed. Each new election brought renewed cries from hill farmers against the railroads and businesses that supported their opponents, but conservatives kept winning by manipulating the black vote. Many whites from both factions felt that, whatever rights blacks had under the Fourteenth Amendment, Mississippi's elections would be run more honestly if they did not vote. A bitter constitutional convention in 1890 seemed to have succeeded in disfranchising blacks, but poor whites soon learned that Delta blacks were still voting and that many of their own number had been disfranchised by the new literacy requirements.

It was not until 1902, a quarter-century after Reconstruction, that poor whites finally managed to assert themselves. Their tool was a new law requiring party primaries. Since membership in the Democratic party, the state's only functional party, could be restricted to whites, blacks were at last effectively disfranchised; tragic for blacks, that development meant liberation for hill farmers, who were free now from the manipulated votes of blacks in the Delta. But not even the most radical agrarian knew how profound those changes were likely to be. The new law had produced a new breed of politicians, one editor lamented. "The men who can shake hands best . . . and tell the funniest stories, have the best chance to win in this state."[6]

James Kimble Vardaman was a Lamar turned inside-out. Delta aristocrat William Alexander Percy, whose father led the conservative opposition, remembered Vardaman as "a kindly, vain demagogue unable to think";[7] but to poor whites he was the "White Chief" who would lead them out of the wilderness. Before black disfranchisement, aristocrats had wheedled votes from poor whites by reminding them that blacks might combine with some white faction to outvote them; with blacks out of the picture, the fears of poor whites centered on blacks themselves and on those aristocrats who employed and sometimes protected them. Vardaman knew how to marshal those fears, and his emergence as a political power meant that for Mississippi's aristocrats the wheel had come full cycle: now it was their opponents who were winning by running against blacks. William Faulkner was six years old when the White Chief rode into power on a wave of poor-white racism. As Faulkner grew older, he must have grown increasingly aware that Vardaman's victory was a pivotal event in his life and in the lives of everyone around him. Since Reconstruction the "ignorant" and the "educated" of Mississippi's white electorate had been locked in a deadly struggle with blacks as the scapegoats, and now the "ignorant" were winning. For Faulkner, as for Du Bois, that was important, for the Falkner family's way of thinking about itself had emerged from the rough and tumble of Mississippi politics.

II

William Clark Falkner (1825–89) was a man of many faces. Sometimes he behaved like one of Horatio Alger's rags-to-riches heroes. A youthful runaway who arrived penniless in the Mississippi hill country in 1840, he remained to acquire land, slaves, and a fortune of $50,000. When the Civil War came, he enlisted early and became a hero at Bull Run; later he became commander of a band of partisan guerrillas. Falkner's feats of military heroism identified the central tension of his career; as his reputation grew, he seemed to be placing between himself and his anomalous origins the persona of the Cavalier aristocrat. After Bull Run Falkner, elected by his men to the rank of colonel, gratefully turned down a promotion to brigadier to remain with them—only to find himself deposed in the next military election. That 1862 election set a pattern that would remain with Falkner: he seemed always to be struggling up to some grand eminence, then getting shot down.

After the war Falkner allied himself with the new businessmen-aristocrats who rallied around Lamar.[8] In 1871, with Ripley businessman Dick Thurmond and others, he began to build a railroad through the Mississippi hill country; by 1888 Thurmond and the others were out of the picture, and Falkner was sole owner of the line. For many men, that success might have stood as a crowning achievement, but Falkner had his mind on other things as well. Always interested in literature, he had written poems and plays; in 1881 his novel *The White Rose of Memphis* sold 10,000 copies. Although Falkner had never committed himself seriously to politics,[9] by 1889 political fame was the only kind that had escaped him; he announced for the legislature that year and was elected handily.[10] With this election Falkner seemed at last to have achieved the acceptance of the people from whom he had sprung. In 1862 a military election had separated Falkner from those people; now another election had sealed that bond again, and it scarcely seemed to matter that Thurmond, whom Falkner had edged out of the railroad, might be gunning for him.

For the colonel's great-grandson and for the rest of the family that he left behind, Falkner's murder by Thurmond was an event that altered the courses of their lives for generations. Colonel Falkner had achieved a different kind of immortality than he had hoped for; instead of breaking the pattern that had shaped his career, he had bequeathed it, as he never could have in life, to his survivors. To young William Faulkner, born eight years after his great-grandfather's death, the Old Colonel could never be more than myth; but what *was* the myth? Had his ancestor been the penniless waif who had risen through hard work to culture, affluence, and acclaim? Or had he been the scornful Cavalier, rightly rejected by the people he had set himself above? If Falkner had served in the legislature of 1889 and if he had struggled through the constitutional convention of 1890, the answers to those questions might have been clear. Now they could never be, and that was Thurmond's revenge: the Falkners would never know whether they were Cavaliers or rednecks, and not knowing would affect all of them profoundly.

That was why, for William Faulkner and his brothers Jack, John, and Dean, the Old Colonel's son John Wesley Thompson Falkner (1848–1922) assumed a special importance. The Falkner boys would know their great-grandfather only through family legend; but their grandfather, whom they called the Young Colonel, was a part of their everyday lives, and he had come to represent everything the Old Colonel's death had taken from them. "He loved his family and his country," Jack Falkner reminisced sixty years later, "the latter meaning, of course, the South"; to Jack it was important that his grandfather "had been born in 1848 and therefore had a clear recollection of the good life of his kind in our state before the war."[11] After the fighting ended J. W. T. Falkner had moved easily into the life-style of the postwar aristocrats, by turns a railroad man, a lawyer, a politician, and a banker. Before his father's murder he was already living in Oxford, home of Lamar, with whom he aligned himself politically.[12] By 1891, succeeding where his father had failed, he was seated in the legislature; by 1895 he was in the Mississippi Senate, by 1899 Lafayette County attorney,[13] and his career had begun to look like a primer of the new businessman-aristocrat's way of life. But the tensions that had shaped the Old Colonel's career were very much alive in his son. The Young Colonel could hardly have been enthusiastic about Vardaman, enemy of railroads and corporations[14]—particularly after he was swept out of office in the wake of the Vardaman landslide of 1903; but he was also a politician. When Vardaman backers in Oxford held a rally for the White Chief in 1910, the president of the newly formed Vardaman club was none other than J. W. T. Falkner.[15]

For thirteen-year-old Bill Faulkner, trying to understand the family heritage must have been frustrating; each new generation of Falkners seemed destined to play some tragic new variation on the Cavalier-redneck theme. The Old Colonel, who had begun with nothing, had made himself an aristocrat and had been killed when he refused to defend himself like one. Now the Young Colonel had been swept out of office by the poor whites and, because he had, had chosen to align himself with them. For Faulkner, that alignment would be important; on the threshold of his puberty, his grandfather had involved the family in a bitter recycling of the Old Colonel's tragic heritage that would last well into his young manhood.

He would later remember Lee M. Russell as a "countryman" his "grandfather had made a lawyer of, . . . who became governor of Mississippi."[16] Russell, a poor white who had thrust his way up from poverty, would be political mentor to J. W. T. Falkner, Jr., soon to become a member of the law firm. When Vardaman associate Theodore G. Bilbo was elected governor in 1915, Russell was the choice for lieutenant governor, and the position of J. W. T. Falkner, Jr., seemed unassailable: since Mississippi governors could not succeed themselves, his law partner was now heir-apparent of the poor-white faction for the governorship. Four years later J. W. T., Jr., was Russell's Lafayette County campaign manager, and the man his father had "made a lawyer of" was in the governor's mansion; Oxford greeted them with a torchlight victory parade, and the Young Colonel's son presided over the ceremonies at the square as toastmaster.[17] For the Falkners, the victory seemed to open up many things. After

Russell became governor, Murry C. Falkner, William's father, was made assistant secretary of the University of Mississippi; Russell rewarded his campaign manager in 1922 by appointing him to an unexpired judgeship on the Mississippi Third District court.

There were ironies in the Falkners' new eminence. Fellow townsmen remembered that their political allegiances had not always been with the poor; and the Young Colonel, a loyal Vardaman backer who had no love for Bilbo, seemed to have trouble remembering that, too. Falkner never accepted his law partner socially; even after Russell was lieutenant governor, he was not invited to the Falkner home. "Sir, our relations are business and political, not social," the Young Colonel said on one occasion when his associate came to call—and slammed the door in his face.[18] J. W. T. Falkner may have had concerns other than social ones on his mind; there was always the implication that anyone in politics was in it for money, and there were neighbors enough to remind the Falkners of such matters. After Russell was elected governor, one Oxford lady called to the Falkner boys as they passed her house, "I reckon you Falkners are hurrying to get on the bandwagon." Young Bill had a quick answer. "No Ma'am," he said. "We're jumping down so you folks who didn't vote for him can climb on quick."[19] But the meaning of this and similar remarks must have been clear to the Falkners since the Young Colonel began to back Vardaman.

William Faulkner had ample opportunity to witness those ambiguities; his brother Jack was soon working in the law office, and, when their uncle began his campaign for re-election to the district court, Bill was often his chauffeur.[20] Oxford druggist Mac Reed remembered the time Judge Falkner asked him to go to Chickasaw County to talk to voters on his behalf, and Bill came along as the driver; the young man seemed indifferent to what he was doing, and Reed was forced to an embarrassing conclusion: "William was not at all interested in politics, even on behalf of his uncle."[21] Faulkner may have been bored, but he may have been learning, too; his uncle, an effective campaign manager, simply was not the kind of candidate with whom hill farmers could live. Russell himself was not long in discovering the hamartia of Mississippi politics. In 1922 a former secretary had named him as correspondent in a paternity suit. Russell had not only gotten her pregnant, she claimed, but also he had forced her to have an abortion—and as if the governor was not star enough for the trial, she brazenly subpoenaed his predecessor, Bilbo, as a witness. When Bilbo refused to appear, Russell was acquitted, but his career was over. In Mississippi in 1922 only people like Bilbo, for whom scandals seemed a way of life, appeared to stay afloat.

To twenty-five-year-old Faulkner, on the threshold of his career as a novelist, the episode must have spoken volumes. His family's patrician image had some cracks in it, and Russell's fall illustrated most of them. The Old Colonel had been killed when it seemed the people had finally accepted him; the Young Colonel, thrown out of office in a voters' rebellion, had aligned himself with the people but never accepted *them*; Faulkner's uncle, who had inherited that ambiguous commitment, had found the people ungrateful not only to himself but to those like Russell who were their own kind. No wonder Reed thought Faulkner was "not at all interested in politics"; every time the family got involved in politics, something happened. A natural response for an ambitious young writer would have been to mutter, "A plague on both your houses!"—and retreat into his books. But on a deeper level that, too, was impossible.

III

Vardaman and Bilbo might be demagogues, but they had been responsible governors in other ways, and the few aristocrats who voted for them could point to genuine reforms; the trouble was that to benefit from those reforms they also had to ignore the politicians' ideas on race. To outsiders, Delta and hill country sounded alike on race; Vardaman spoke for most white Mississippians when he declared blacks were "inherently unmoral, ignorant and superstitious, with a congenital tendency to crime, incapable unalterably of understanding the meaning of free government."[22] But the *quality* of the racism of Mississippi's white factions, as Du Bois knew, was in many ways different.

Secure in their inherited status, aristocrats looked complacently down on both blacks and hill folk. To aristocrats, each group appeared hopeless when it came to manners, to morals, to *savoir faire*; and, if aristocrats brought *any* plebeians into their world, they were likely to be blacks who had learned plantation manners as household servants. The aristocrats' status sometimes made them think of themselves as tolerant toward blacks. They liked to think of blacks as their feudal responsibilities, not as their enemies; their heritage of *noblesse oblige* told them honor was at stake if they failed to protect what they looked on as their peasantry. The conditions of plantation life fostered this illusion. Aristocrats lived with blacks intimately. Unlike the hill folk, aristocrats were raised by black servants, played with their children, and were sometimes suckled by black "mammies." Field hands, moreover, were investments, a fact that afforded aristocrats a strong motive for protecting them even after they had been "freed" and called themselves sharecroppers. As the plantation gradually became a sentimental liability, the change in the planters' economic base to business did little to alter such motives; cheap labor was an asset, and in a declining economy assets were hard to come by.[23]

Percy, whose father led the patrician opposition to Vardaman, felt he appreciated blacks. This, he thought, was possible because he understood his own limitations in dealing with them. Although black and white Mississippians "believe they have an innate and miraculous understanding of one another," he wrote in 1941, "the sober fact" was that "we understand one another not at all." He had found that "just about the time our proximity appears most harmonious something happens . . . and to our astonishment we sense a barrier between." What was worse, "the barrier is of glass; you can't see it, you only strike it." Barriers or not, Percy thought he was aware of "the Negro's excellences." "The Southern Negro has the most beautiful manners in the world," Percy insisted, and he had a formidable list of other excellences: "his charm, his humor, his patience, his exquisite sensibilities, his kindness to his own poor, his devotion and sweetness to all children, . . . his poetry of feeling and expression, his unique tactual medieval faith, his songs more filled with humility and heart-break than Schubert's or Brahms's."[24]

Who could fail to be moved by such graces? "I want with all my heart to help him," Percy said, but there were problems—for example, inherited racial characteristics. "Poor wretches! For a thousand years and more they had been trained in tribal barbarism, for a hundred and more in slavery"; the result was "a race which at its present stage of development is inferior in character and intellect to our own." Percy thought he could see the proof all around him. "None of them feels that work *per se* is good; it is only a means to idleness." That was because "the American Negro is interested neither in the past nor in the future. . . . He neither remembers nor plans." What was worse, blacks seemed to have no moral sense. "Murder, thieving, lying, violence—I sometimes suspect the Negro doesn't regard these as crimes or sins." But there was a beguiling innocence even in their violence: "Negroes are as charming after as before a crime." That was why that "committing criminal acts, they seem never to be criminals." It was the duty of responsible whites to protect these people from themselves. "All white families expend a large amount of time, money, and emotion in preventing the criminals they employ from receiving their legal deserts." Thus, despite the invisible barriers, Percy thought he knew how to treat the black. He was "our brother," he insisted; but he was "a younger brother, not adult, not disciplined, but tragic, pitiful, and lovable." The aristocrat had to "act as his brother and be patient."[25]

Patricians like Percy hid their racism from themselves under a smoke-screen of benign paternalism. They needed that irresponsible "younger brother" more than he needed them; without an appropriately irresponsible working class, there was little need for a responsible aristocracy. And if aristocrats like Percy were appalled when poor whites called their "younger brother" a beast and claimed he was dangerous, it was at least partly for the same reason: an elite class was a travesty if it could not control its work force. A Percy could admit blacks were dangerous to each other, but he could never admit a danger to anyone else, and he was ready to place the blame for all racial unrest on "the lawlessness and hoodlumism of our uneducated whites."[26] For Percy the race problem was the poor-white problem.

Percy found it hard to be objective about poor whites; he was dismayed as his father campaigned "for fair treatment of the Negro" before one hostile audience. "They were the sort of people that lynch Negroes, that mistake hoodlumism for wit, and cunning for intelligence, that attend revivals and fight and fornicate in the bushes afterwards. They were undiluted Anglo-Saxons. They were the sovereign voter. It was so horrible it seemed unreal." He theorized that because such people were "intellectually and spiritually . . . inferior" to the black, they thought they had to "do something to him to prove . . . their superiority." It was these people who were responsible for "the disgraceful riots and lynchings," Percy insisted, and he was sure that only a return to patrician control would solve the problem. "Our fight to protect the Negro is merely part of our fight for decency in politics, for law enforcement."[27]

Poor whites, predictably, took issue on that point; to them the race problem was the aristocrat problem. The hatred of poor whites for blacks recalled their grievances against the antebellum slaveholders who were responsible for inundating the country with cheap black labor in the first place. Since antebellum days aristocrats had monopolized the state's best land, and poor whites looked on the blacks who worked it almost with envy. Some blacks, they knew, had other advantages, too: household slaves enjoyed an acceptance and a protection to which few poor whites could aspire. Like blacks, poor whites had looked to Reconstruction for redress for their grievances. They had been outraged when with the Redemption life seemed to be reverting to its prewar state, and, as the cotton market sank, they found themselves reduced to the same level as their black neighbors. There were cultural motives for their outrage, too. The less prescriptive moral attitudes of some blacks shocked fundamentalist sensibilities, and the "primitive" Christianity practiced by many blacks scandalized them; whites tended to connect the planter's permissiveness in these matters with his own bland indifference to militant Protestantism. For the poor white, black and aristocrat thus blended into a cluster of hated stereotypes.

Vardaman's racism, like Percy's, found its origin in his notions of black irresponsibility; but in turn-of-the-century Mississippi it was not what one claimed to see, it was how one reacted. Where Percy saw an irresponsible "younger brother," Vardaman saw a dangerous beast. The black, he proclaimed, was a "lazy, lying, lustful animal," whose nature "resembles the hog's."[28] In one speech he asked his audience to imagine the rape of two white women by a black "fiend"—"a young buck negro": "Buck passes down the street. He spies a little cottage; . . . he finds a mother and daughter there, the husband and brother out in the field at work. The beast goes in and commits that crime which forever blasts the peace, the purity, and the happiness of that home. [The constable] . . . may possibly arrest him. . . . But he has not brought back the love and light and purity of that home."[29] There was an evangelism in Vardaman's manner, as though he were exhorting his constituents to a war against sin itself. Sometimes he seemed to be calling for genocide: "We would be justified," he claimed, "in slaughtering every Ethiop on the earth to preserve unsullied the honor of one Caucasian home." If a man saw a wolf, he ought not to wait "to find if it will kill sheep before disposing of it."[30]

It was the poor whites' fears that made them feel more secure listening to Vardaman's harangues; and, whatever the sources of those fears, they set the tone of an era. Young Faulkner had been born into a society which had been bombarded for more than a decade by racist propaganda that, from the time he was old enough to have some understanding of it, had reached a scope and intensity almost beyond belief. Vardaman's Jackson newspaper *The Issue* ran weekly articles under headings such as "Sexual Crimes Among Southern Negroes Scientifically Considered," "Cannibalism in Hayti," and "The Negro a Different Kind of Flesh." Some of these were by supposedly responsible academics; others were by blatant racists like ex-Baptist minister Thomas Dixon, Jr.[31] Faulkner's personal library contained a copy of Dixon's *The Clansman,* given to him by his first-grade teacher, Annie Chandler.[32] In this "romance of the Ku Klux clan," as Dixon called it, Faulkner could have read how Republicans had orchestrated the Reconstruction solely to enhance their political power at the expense of the white South. He could have read of an unruly black South Carolina legislature, presided over by a "mulatto," voting itself money from the state treasury. And he could have read of a black army captain with "thick lips," which "were drawn upward in an ugly leer," whose "sinister bead-eyes gleamed like a gorilla's"[33]—and could have followed this creature as he raped a white mother and her daughter, who responded to their defilement with a suicide pact at "Lover's Leap."

If Faulkner was slow in turning the pages of the novel his teacher had given him, he had only to attend the Oxford Opera House, a Falkner family institution managed by Russell. One attraction in 1908 was Dixon's stage version of *The Clansman* — later filmed as *Birth of a Nation* — to which customers were lured by a large advertisement picturing riders wearing hoods.[34] Russell had reason to think Dixon's drama would draw well, for seven weeks earlier an incident had occurred that to many of Oxford's citizens made it seem timely. Mrs. Mattie McMillan lived in a small house north of town with a teen-aged daughter and two smaller children; her husband, a prisoner in the county jail, had sent a message to her by a black trustee, Nelse Patton. When Patton arrived, reported the Lafayette County *Press*, Mrs. McMillan went for her gun, convinced that this "desperado" meant to rape her. The *Press* related what allegedly followed: "One of the coldest blooded murders and most brutal crimes known to the criminal calendar was perputrated one mile north of town yesterday morning . . . , when a black brute of unsavory reputation by the name of Nelse Patton attacked . . . a respected white woman, with a razor, cutting her throat from ear to ear and causing almost instant death."

Eleven-year-old Faulkner would not have had to read the *Press* to find out what happened. Deputy Sheriff Linburn Cullen, father of Bill's classmate Hal, had forbidden his two older sons John and Jenks to join the posse he was helping to organize; ignoring their father, they armed themselves with shotguns and, guessing that Patton would head for Toby Tubby Bottom, found him before the posse. There, reported the *Press*, Patton "suddenly ran amuck of Johnny Cullen," and the boy responded like a deputy's son: "Seeing the negro coming towards him, . . . [young Cullen] called a halt, but the negro paid no attention to the command and the boy let him have a load of No. 5 shot in the chest, which slackened his speed but did not stop him. The boy gave him another charge in the left arm and side which stopped him. The negro was at once surrounded by his pursuers and gladly gave up. . . . Being weak from loss of blood, the brute was put on a horse and hurried to jail."

There was little doubt that Patton had killed Mrs. McMillan; a piece of metal taken from one of her vertebrae fitted a gap in his razor. But few of the nearly 2,000 people who crowded around the square seemed to think he might have done it to keep her from shooting him; most seemed to agree with former U.S. Senator W. V. Sullivan that Patton had done what poor whites expected blacks to do to their women. Responsible Oxford citizens took turns pleading with the crowd to go home, but Sullivan, whose nomination J. W. T. Falkner had seconded eight years earlier, began to harangue them to go after Patton. As the crowd moved toward the jail, Sullivan gave his pistol to a deputy. "Shoot Patton," he said, "and shoot to kill." When they broke in, the prisoner felled the first three with a railing from his bed; then, reported the *Press*, "26 pistol shots vibrated throughout the corridors of the solid old jail, and when the smoke cleared away the limp and lifeless body of the brute told the story. The body was hustled down stairs to terra-firma, the rope was produced, . . . and the drag to the court house yard began."

The *Press* failed to report that by this time Patton's corpse had been stripped of its clothes, castrated, and mutilated, but it did not shrink from telling its readers what happened next: "This morning the passerby saw the lifeless body of a negro suspended from a tree—it told the tale, that . . . the public had done their duty."[35] The coroner's jury ruled that the victim "came to his death from gunshot or pistol wounds inflicted by parties to us unknown." Sullivan sought no such anonymity. "I led the mob which lynched Nelse Patton and I am proud of it," he told the Associated Press. "I wouldn't mind standing the consequences any time for lynching a man who cut a white woman's throat."[36]

And such people, it appeared, were the only ones who ever were lynched. In Mississippi in 1908 the collective imagination of poor whites orchestrated such realities, and the lynching of black "rapists" occurred with machine-like frequency; such tales were standard items of contemporary folklore, inherited by every southern child. Sullivan was claiming more credit than he deserved; any of the twenty-six shots fired at Patton could have killed him, and all of the citizens of Oxford except the few who had tried to disperse the mob had a share in what had happened. No wonder Percy had thought the crowd that heckled his father "was so horrible it seemed unreal"; the Sullivans and Vardamans were of a very different order from the Percys or—for all the Young Colonel's political allegiances—the Falkners. For any sensitive young aristocrat coming of age in turn-of-the-century Mississippi, there was never any question about which reality to choose; events like the Patton lynching validated family pressures so that any other choice was improbable.

The problem was that there were only two sides to choose *from*. If a young aristocrat was shocked by redneck atrocities like the Patton lynching, he might, for that very reason, be even less likely to question the bland racism of paternalists like the Percys, who claimed to be working "for fair treatment of the Negro." For eleven-year-old Faulkner, at least, the choice had already been made. Sixteen years later, when he took time out from writing *Soldiers' Pay* to put together a sketch for the New Orleans *Times Picayune*, it was clear enough what it had been. "Sunset" (1925) began with a fictional news story that sounded like a parody of the *Press* account of the Patton lynching.

BLACK DESPERADO SLAIN

> *The Negro who has terrorized this locality for two days, killing three men, two whites and a negro, was killed last night with machine gun fire by a detachment of the -th Regiment, State National Guard. . . . No reason has been ascertained for the black's running amuck, though it is believed he was insane.*[37]

The *Press*'s reporter and Faulkner's spoke the same language; each called his black a "desperado" and claimed he had run "amuck." But newspaper accounts, Faulkner seemed to be saying, did not always tell the whole truth, and, as this desperado's story unfolded, it became clear that, whatever his sins, he was no black rapist (*NOS*, 80).

"Ah wants to go back home," he told a startled Canal Street policeman the day before he was shot, "whar de preacher say us come from." He had left "Mist' Bob and de fambly and his niggers" to go back to Africa, blandly assuming New Orleans must be the first stop; but he had no idea where Africa was, and he was too frightened by the traffic to cross the street. "It took him two days to come from Carrolton Avenue to Canal Street." At the docks he boarded a boat and asked a white sailor if its destination was Africa; when he was told it was going to Natchez, he replied, "Dat suit me all right, jes' so she pass Af'ica." The dumbfounded sailor told the old man to go to work with the "other boys," took the four dollars he offered for passage, and, when the boat was some miles up the Louisiana shore, told him that "Africa is about a mile across them fields yonder" (*NOS*, 82, 80, 85, 86).

To this point Faulkner's whites had seen nothing to fear in the shotgun the old man carried, but now it got him in trouble. On shore in the dark he fired at an animal he thought must be a lion; when the shot drew a swarm of angry, French-speaking farmers, he fired at them, too, convinced they must be "savages, . . . folks that eat you"—and he shot three of them. Wounded and delirious, hunted down like an animal, Faulkner's first black hero died with no visions of Africa's green hills: "He was at home, working in the fields," Faulkner wrote, and presently the old man saw "supper in the pot." And that was how he died, dreaming that "tomorrow he'll be home, with Mr. Bob to curse him in his gentle voice, and regular folks to work and laugh and talk with." Africa was a dream; but heaven for Faulkner's first black hero was "Mist' Bob and de fambly and his niggers."

"Sunset" looked like a plea for understanding; Faulkner seemed to be asking readers to look behind the brute stereotype. But finding the human being behind the stereotype was not so easy for Faulkner. His old man was no animal, but he did not seem quite human either; he talked like a minstrel side-man, he was self-effacing as Uncle Tom himself, and he was so out of touch with reality that whites had to look after him. Faulkner had stripped away one black mask, the mask of the brute, but under it he had found only the aristocrat's "younger brother," and if there was a human being anywhere around, he was not visible.[38]

Du Bois was right: there were many Souths, and the South of "the sons of the masters" was sometimes a more hopeful South than that of Vardaman. But, as Du Bois also knew well, that South had its limits, too; the trouble was that the sons of the masters wanted only *one* South, and, although in 1903 that South might appear safer, it looked all too familiar. "Sunset," like other stories of its time, showed what South that was. Faulkner was pleading for more than his doddering black hero in "Sunset"; he was pleading for aristocrats and the world of his youth. What "Sunset" showed was how southern society, saddled with hostile non-aristocratic whites like Faulkner's sailor and French farmers, needed Mr. Bob and his plantation. Faulkner had dug in precisely where his class had dug in against the "revolt of the rednecks." He was defending the South he thought the Old Colonel had left him, and he was doing that the same way his class had done it since Lamar and the Redeemers: trotting out appropriately hostile outsiders and irresponsible blacks to show why that South had to be preserved.

3

A Visit to a Familiar Place

I

"You're a country boy," Ohio novelist Sherwood Anderson told the great-grandson of Colonel W. C. Falkner, "all you know is that little patch up there in Mississippi where you got started from." To Anderson it did not matter where a writer was *from;* like the rest of America, Faulkner's area was so important that if you were to "pull it out [of the country's structure], as little and unknown as it is, . . . the whole thing will collapse, like when you prize a brick out of a wall." Anderson thought his friend ought to go home and write about what he knew about because Mississippi was "all right too. It's America too."[1]

Anderson's friend seemed to be taking his advice. He *was* writing about a small southern town, although he had placed it in Georgia, not Mississippi; he had given his town the Cavalier name of Charlestown; and his hero, ex-British flier Donald Mahon, was an aristocrat, the son of the local rector.[2] But Faulkner seemed to have other things on his mind. He walked with a limp and carried a cane to steady himself: he had been wounded in the war, he said, and had a silver plate in his head to prove it. And the novel he was writing was not so much about Charlestown as about the town's reaction to Donald, who had come home fatally wounded, a hideous scar across his forehead. Anderson, who helped Faulkner with his manuscript and asked his own publisher to read it, did not question his friend's war experiences; he did question why Faulkner wanted to write about war instead of about Mississippi.[3]

Colonel Falkner might have understood. In the Old Colonel's South one enlisted when war came and won honors or died trying, and his great-grandsons had been quick to volunteer in World War I. Fifteen-year-old John Faulkner convinced a recruiting officer he was eighteen and his father had to call on the Mississippi adjutant general to stop him; nineteen-year-old Jack Falkner joined the Marines; and the tastes of the older brother, though more selective, were fully as Cavalier—William Faulkner wanted to be a flier.[4] World War I fliers were a special breed: the last, perhaps, to find a glamour in war like that the Old Colonel's generation had claimed. But Faulkner soon discovered that glamour and flying did not always mix. Rejected as an Army flier because he was not tall enough, he joined the Royal Air Force, but from the Cavalier standpoint, at least, his career was an embarrassment: Faulkner spent his war training in Canada. It was the middle Falkner brother who came home limping, wearing a head scar, if not a silver plate, and wrapped in laurels like the Old Colonel's.[5]

What Anderson had failed to understand was that in writing about war Faulkner was indeed writing about his family and "that little patch up there in Mississippi." Donald Mahon's war was the war Faulkner had tried, but failed, to get into; Donald had fought it in the air, where the glamour was. His homecoming had been more like Jack Falkner's—the homecoming, at least, that all the Falkners had feared Jack would have: like Jack, Donald had been given up for dead, and, when he finally arrived back home, he was fatally wounded, ready to sink into a terminal coma. But Faulkner had a great deal more in mind in *Soldiers' Pay* than working off his fantasies about the war, and he had already made a decision to insure that his novel would be more than another wallow in Lost-Generation *Weltschmerz.*

When Faulkner first met Anderson in 1925, the Ohio novelist had been working on *Dark Laughter.* Anderson had gotten that title, he said later, from the way he used blacks in the novel; he had set out to use "the mysterious, detached laughter of the blacks" to suggest a world of experience not available to inhibited whites: "dark, earthy laughter—the Negro, the earth, and the river—that suggests the title."[6] *Soldiers' Pay* featured a chorus of black characters who functioned strikingly like Anderson's,[7] but there were more of them; their presence pervaded the novel and gave it a special meaning not found in Anderson's. It was one of those decisions that change careers. Indigenous as the pursuit of military glamour might be to his Mississippi heritage, writing about that pursuit might have allowed Faulkner to postpone any confrontation with Mississippi; but Faulkner's chorus of blacks placed him in a situation where he could scarcely ignore it.

The first of these blacks, a railroad porter, could have been straight out of Murry Falkner's household. "Uncle Ned" Barnett, John Faulkner wrote years later, had been "a slave belonging to my great-grandfather." After the Old Colonel's death Ned had gone to Oxford, where he worked first for the Young Colonel, then for Mr. Murry;[8] later he would attach himself to Faulkner. Ned referred to all his patrons as "Master." Like the

Old Colonel himself, he was a man of many faces. Sometimes Ned was bizarre. Clothes were his foible; he kept two trunks, Faulkner remembered, which bulged with cast-off garments from four generations of patrons, and on weekends he ventured forth in sartorial splendor.[9] Ned wore a necktie even while chopping wood or milking the cow.[10] But Ned's comic persona masked a feisty independence. "He wanted to take part in everything and boss all of it," John remembered.[11] Not above conning the Falkners when he could, Ned was also judgmental, measuring all of them by the more heroic standards of the Old Colonel. "He is a cantankerous old man," Faulkner said once, "who approves of nothing I do."[12]

Faulkner's porter was like a Ned Barnett forced to make his way outside the South. An old-fashioned man who had not forgotten the South of his youth, he sensed Donald's patrician breeding and took the wounded flier under his wing: he was "deftly officious," Faulkner wrote, "including . . . [Donald and his party] impartially in his activities, like Fate." When a veteran, Joe Gilligan, tried to help, the porter seemed hostile until he learned that this northerner meant to act responsibly; then he elevated Joe to Mahon's status. When Alabama-born Margaret Powers attached herself to the group, he confided, "I'm from Gawgia"—and added, "We got to look out for our own folks, ain't we?" (*SP*, 27, 34). Faulkner's porter was more complex than the old man in "Sunset"; he was a survivor, no helpless "younger brother"; as a railroad man, he did not *live* anywhere, and in his own way he was experiencing the alienation that Faulkner's whites faced. But the portrait, almost gratuitously inserted, suggested a point that seemed irrelevant in this Lost-Generation context: his attachment to the old ways combined with this very complexity and independence to argue, by implication, the familiar sentiment that blacks preferred servitude in the South to independence elsewhere.

"Uncle Ned" Barnett had been a family institution since slavery; Caroline Barr, hired by Mr. Murry in 1902,[13] had become one. Mammy "was not considered a servant by the family," Jack Falkner wrote; like other "mammies" of the era, she "was as much a member of the family as any of the rest of us." This tiny, garrulous woman presided over the Falkner boys' youth with bustling authority and a scathing tongue. Like Ned, Mammy had her comic persona, and the family laughed affectionately at her foibles. When Jack was wounded, Mammy wanted to know the whereabouts of "dis here France where de Germs is shot Jackie"; despite Miss Maud's efforts to explain, she remained convinced that "hit sho'ly be's jest t'other side uv Arkinsaw!" But, like Ned, she held no one in awe. "I have seen her on a street corner," Jack recalled, "completely at ease, talking to some business or professional leaders of the town, and when I returned . . . I would see her . . . [talking] with some cooks and maids from nearby homes."[14] When Mr. Murry died, Mammy would be passed on to Faulkner as Ned was; she died at age 100, a revered figure who was central to the family's self-image.

Faulkner could not resist creating a mammy for his wounded flyer in *Soldiers' Pay.* Back home in Charlestown, the comatose Mahon dreamed away his days in a yard chair, and it was here that Mammy Callie Nelson appeared with her grandson, a gawking youth in a private's uniform. "Hush yo' mouf, Loosh," she said, "it'll be a po' day in de mawnin' when my baby don't wanter see his ole Cal'line." With that she addressed Mahon. "Donald, Mist' Donald honey, . . . here yo' mammy come ter you." When the flyer did not respond, she asked, "Don't you know who dis is? Dis' yo' Callie whut use ter put you ter bed, honey." After examining Mahon, she concluded, "Lawd, de white folks done ruint you" (*SP,* 170).

Not everything in Callie Nelson's portrait was quite so bizarre. She exuded some of Caroline Barr's garrulous strength, for example; no longer employed by Rector Mahon, she lived in proud independence. That pointed toward the novel's central theme: if "white folks" had "ruint" Donald, Mammy Callie's comment reinforced the polarity Faulkner was setting up between alienated whites and blacks who lived earthier, simpler lives. But Donald's mammy, like Faulkner's porter, seemed more like characters out of Thomas Nelson Page than out of a novel of Lost-Generation alienation; Callie Nelson spoke the caricatured dialect of the minstrel show, and she behaved toward Donald more like a clown than a foster mother. Worse yet, the portrait seemed largely gratuitous: Mammy Callie was never mentioned when the novel flashed back to Donald's childhood, and she was mentioned only once—briefly—during the remainder of the novel.

Brief as portraits like Callie Nelson and the porter were, they bent Faulkner's novel temporarily out of shape. Faulkner seemed drawn to them compulsively. It was as though he were writing two novels, one about Lost-Generation whites and the other about blacks and paternalism, and often the second intruded on the first, obscuring its tragedy with pejorative clichés and unexplained digressions. When one black smiled, "white teeth were like a suddenly opened piano." A "negro driver's head," Faulkner reported, "was round as a capped cannon-ball." Blacks surrounded the square in "stolid coagulation"; passing whites were "lapped, surrounded, submerged by the frank odor of unwashed negroes" (*SP,* 16, 207, 146, 144).[15] Lines like these showed how far Faulkner had to go before coming to grips with the meaning of his black chorus;

fortunately for *Soldiers' Pay*, however, he had discovered other ways of thinking about blacks that did not originate in Mississippi.

Anderson, talking about the "dark, earthy laughter" he associated with blacks, was drawing on an image of black life emanating out of Harlem, not Dixie. Used by black as well as white writers in the 1920s, this image was in most ways as stereotyped as those it displaced. The shadows of Pablo Picasso and Georges Braque hovered over the ghetto. Blacks were said to possess a special kind of beauty that resembled African carvings; they were supposed to be free of inhibitions and easily swept up by passion—to possess an easy sexuality unavailable to whites, to be more potent. The (to whites) bizarre clothing favored by some working-class blacks was no longer derided; now it was quaint, mocking the drab fashions of Main Street. And being "primitive," whites began to suspect, meant access to a ready spirituality inaccessible elsewhere; to some white intellectuals, it explained why blacks were spontaneous musicians who could produce America's greatest religious music—the spirituals.[16]

Writing about blacks as exotics made Faulkner more subtle. Dark laughter reverberated throughout his novel. Those "unwashed negroes" who surrounded the square, Faulkner reported, possessed a "careless, ready laughter." Gilligan, walking through a black neighborhood, heard "soft meaningless laughter" from the houses. The tragedy of slavery always seemed to be a part of such laughter; it was "somehow filled with all the old despairs of time and breath." But most of all, Faulkner's blacks seemed to express their condition in an elemental music. The novel ended with Gilligan, who had just lost Margaret, walking the dark streets with the recently bereaved Rector Mahon. As they passed a black church, they heard voices: "the crooning submerged passion of the dark race." The nature of that passion was elusive: "It was nothing, it was everything." But in it, Christianity was transformed into a soaring paganism: "it swelled to an ecstasy, taking the white man's words as readily as it took his remote God and made a personal Father of Him." And that ecstasy was transcendent: "a clear soprano of women's voices" rose "like a flight of gold and heavenly birds." As the two men watched, the "shabby church" became "beautiful with a mellow longing, passionate and sad." When the singing ended, they knew that its beauty was "inevitable," too, along "with to-morrow and sweat, with sex and death and damnation" (*SP*, 144, 312, 319).

Faulkner was learning about blacks and about the masking rituals of the world of his youth. In "Sunset" he had stripped away the poor-white "beast" mask and found only the aristocrat's "younger brother." In *Soldiers' Pay* he had peeled away the younger brother and found the exotic. And freed thus in some measure from the "invisibility" of black identity, he had been able to create a moving reflection of human suffering and aspiration. Few blacks would be enthusiastic about the pejorative images in *Soldiers' Pay*, but few, too, would fail to be impressed with the young patrician's limited but suggestive insights. And what his novel failed to suggest about the realities of black experience, it almost compensated for through what it suggested about whites; he had learned to use black success as an ironic gauge of white failure. *Soldiers' Pay* thus identified a paradox that would remain with Faulkner for the rest of his career. In his fictional Georgia of 1926 the word was still segregated; yet despite the handicaps such segregation implied, the young artist had produced a moving human tragedy.

Nonetheless, the image of black life in this novel was partisan, and it showed where Faulkner was at this stage of his search for a South. There was a "kinship," black critic Sterling A. Brown wrote years later, between the stereotype of the "contented slave" and the exotic; one was "merely a 'jazzed-up' version of the other," he thought, "with cabarets supplanting cabins, and Harlemized 'blues,' instead of the spirituals and slave reels."[17] To Anderson, who had no interest in the southern past in *Dark Laughter*, that kinship was unimportant; but to Faulkner, it mattered a great deal. The attitudes of Faulkner's class were indigenous parts of his personality, and when he focused his attention on exotics he saw them through the lens of his tradition. It was his inability to see them in other ways that led Faulkner to dilute his images of transcendent black humanity with clichés and side-man dialect. And it was that inability which predestined *Soldiers' Pay* to be less than the moving tragedy of postwar alienation that Faulkner intended.

What *Soldiers' Pay* did show was that Anderson, for all his hyperbole, had been right about his friend. Faulkner might know about more than "that little patch up there in Mississippi," but that little patch was what he wanted to write about. Veterans like Gilligan and Mahon existed for Faulkner only in the present, and when he was writing about them, the vitality of his family's past went untapped. Blacks spoke to him about that past, and his fascination with them in *Soldiers' Pay* made it clear where his heart was. Faulkner was not aware of it, but he was already trying to reclaim that world of his youth that time and the revolt of the rednecks had taken from him.

II

Yet Faulkner seemed determined to put his past behind him. Young Americans of the 1920s went to Paris, and by 1925 Faulkner made that pilgrimage; he was living in the Latin Quarter, surrounded by the artistic community that had charmed his contemporaries. But here a striking reversal occurred. Faulkner seemed content to live alone in Paris, and he never made the literary contacts other young writers were seeking—even though letters from Anderson would have opened those doors. "I knew of Joyce," he recalled later, "and I would go to some effort to go to the cafe that he inhabited to look at him." Still, Faulkner never introduced himself, and Joyce "was the only literary man that I remember seeing in Europe in those days."[18]

That withdrawal, too, appeared to have something to do with being from Mississippi. "It seems to me," he recalled, "that very few Southerners were a part of that group of expatriates of the 'Twenties in Paris.' "[19] Europe, at any rate, seemed to be changing Faulkner's outlook toward aristocrats. Donald Mahon of *Soldiers' Pay* had been a tragic Georgia patrician; Faulkner's new hero, Elmer Hodge, was a poor white,[20] and the story's mood was comic. Donald had been wounded in the glamorous air war; Elmer had acquired his limp when he failed to throw a grenade far enough in practice. Unlike Donald, who had come home after the war to die, Elmer had made his way to Paris to become a painter; and now, as Faulkner's poor-white hero pondered Europe's aristocratic heritage, he seemed to be rejecting the very principle of aristocracy in the twentieth century. That heritage was "an old and splendid thing," Elmer thought; but it had had to be "changed . . . to keep pace with the changing times." Europe's aristocracy existed in "a world in which there was no longer any place for it," and what was left of the heritage was "soiled with handling, useless."[21]

In New Orleans Faulkner had appeared to be searching for an aristocratic hero, even a moribund one like Donald Mahon; in Europe he seemed inclined to separate himself from aristocrats and identify himself with poor whites. But what happened now was even more suggestive. Something about *Elmer* had not satisfied him, Faulkner said later; the story was "funny, but not funny enough," and he gave up on it.[22] Back in the South in 1926 he seemed ready to cut other ties. He began a satirical novel about a group of New Orleans artists and *bon vivants* whose yacht was stranded in the isolated reaches of Lake Pontchartrain. These aesthetes were as intellectually barren as the aristocrats Elmer Hodge encountered; confronted with raw nature, they could only respond with words. "Well, it is a kind of sterility, words," one of them mused. "You begin to substitute words for things and deeds" (*MOS*, 210). If Faulkner had little to say to Paris's expatriates, he now had less to say to his Bohemian friends in America. He was ready to embark on a new journey, and he seemed to know where it would take him. Significantly, he now mapped out not one route but two.

Faulkner, Phil Stone told the readers of the Oxford *Eagle* in 1927, "has been . . . writing two novels." One was about some "typical 'poor white trash,' " and Stone thought it might "be the funniest book anybody ever wrote"; the second was "a tale of the aristocratic, chivalrous and ill-fated Sartoris family"—a family of such charisma that one of their ancestors "was even too reckless for the daring . . . Jeb Stuart."[23] His friend was finally ready to write about Mississippi, and Stone's tone identified the younger man's social outlook: "white trash" were comic; patricians were "ill-fated" and tragic. Faulkner named the first of these manuscripts *Father Abraham*, and after many revisions it became his trilogy *Snopes*; the second he called *Flags in the Dust* and published it as *Sartoris* (1929). Stone's protégé had undergone a moving experience that showed him the direction his career was to take. "Beginning with *Sartoris*," Faulkner said later, "I discovered that my own little postage stamp of native soil was worth writing about." The experience had shown him a method as well as a subject. He had found that "by sublimating the actual into apocryphal I would have complete liberty to use whatever talent I might have to its absolute top." There remained only the struggle to get it all on paper, for he knew "that I would never live long enough to exhaust it."[24]

The characters in Faulkner's white-trash novel resembled James Kimble Vardaman's constituency. As the tale began, Flem Snopes, who had struggled up from poverty, surveyed the streets from the window of a Jefferson, Mississippi, bank. Flem had just been elected its president, and the story recounted the plodding skulduggery that got him there. Like the biblical Abraham, Flem was patriarch of a loosely knit tribe of quarrelsome nomads, who had come to Jefferson from Frenchman's Bend, a settlement southeast of Jefferson "in the hill cradled cane and cypress jungles of [the] Yocona River."[25] Faulkner was no stranger to the Yocona River, which ran through Lafayette County southeast of Oxford; placing Frenchman's Bend there meant he was placing Snopes country not far from Dallas, Mississippi—home of the Young Colonel's law partner, ex-governor Lee M. Russell. The basic stock of Frenchman's Bend were Scotch-Irish sharecroppers who for years had been casually exploited by their squire-like landlord, Uncle Billy Varner. These sharecroppers were rednecks an aristocrat could feel at home with; they struggled with nature, not against their landlords or each other, and they managed to maintain the amenities of civilized life. The Snopeses were latecomers who had begun to crowd the older settlers.[26] They had foresworn the amenities, if indeed they had ever understood them. The Snopeses were determined not to struggle *with* poverty but *out* of it, whatever the human cost. Flem was their leader because he had shown them how to struggle and win. Paternalistic Uncle Billy could exploit his sharecroppers and still treat them humanely on a personal level. White trash like Flem reduced exploitation to a science, and that was why Flem became Faulkner's symbol of the revolt of the rednecks.

But Faulkner was not ready to write Flem's story, and he abandoned *Father Abraham* as he had abandoned *Elmer*. Faulkner's own family was on his mind. He had come to realize that there was a world that he "was already preparing to lose and regret." At twenty-nine he had been "feeling, with the morbidity of the young, . . . that growing old was to be an experience peculiar to myself"; he had begun to wonder if a day would come when he would no longer "react to the simple bread-and-salt of the world" as he had. Suddenly he was anxious "to recreate between the covers of a book the world," to "capture . . . that world and the feeling of it as you'd preserve a kernel or a leaf to indicate the lost forest."[27] The world of his youth was fading in 1927, but Faulkner had resolved to take hold of it before it was gone.

Faulkner had been experiencing familiar problems. Words, his characters in *Mosquitoes* had learned, could produce "a kind of sterility"; Faulkner felt he had "been these 2 years previously under the curse of words." Now he found the "touchstone" for which he had been searching. "I . . . realized that to make . . . [this world] truly evocative it must be personal." That would "preserve my own interest in the writing" and "preserve my belief in the savor of the bread-and-salt." As Faulkner described the experience two years later, the "bread-and-salt" of his youth made him think about blacks. His material had come "out of tales I learned of nigger cooks and stable boys of all ages." These ranged "between one-armed Joby, 18, who taught me to write my name in red ink on the linen duster he wore for some reason we have both forgotten, to old Louvinia who remarked when the stars 'fell' and who called my grandfather and father by their Christian names." Whatever the sources, he had learned to sublimate "the actual into apocryphal." His characters were "created," he asserted, "because they are composed partly from what they were in actual life and partly from what they should have been and were not."[28]

The manuscript began with aging Bayard Sartoris musing over an ancient Toledo blade. This relic had belonged to Aylmer Sartoris, an ancestor who had "followed Henry Plantagenet to Rouen"; there Aylmer had wed "the Provencal lady who had borne him the first Bayard Sartoris."[29] This youth, named for the famous "*Chevalier sans peur et sans reproche*," was to take that blade to Agincourt; a later Sartoris, apparently destitute at the hands of Roundheads, brought it to Virginia. Aylmer Sartoris had found his South in Provence and bred southern

European blood into the veins of his descendants; those descendants, fleeing the Puritans, had found a new South in Virginia.

Faulkner eventually cut all these lines from the novel he submitted in 1927 as *Flags in the Dust,* but in these two fragments—*Father Abraham* and *Flags in the Dust*—he had sketched out the territory in which his search for a South would take place. They were like one novel with parallel plots that told of the refighting of the parliamentary wars of the seventeenth century in the New World. Banished to Mississippi, his Cavaliers would find their banners twice brought low by new bands of Puritans: first by those called Yankees, then by those called Snopeses. They would survive these catastrophes as they had the Interregnum—defeated at last only by their own tragic archaism. Asked in 1958 if his books should be read in "a particular sequence," Faulkner suggested that "a book called *Sartoris*" be read first because it "has the germ of my apocrypha in it."[30] With *Father Abraham,* it contained the germ of every record he would leave behind in his search for a South.

III

By the time Faulkner submitted *Flags in the Dust,* the geopolitics of Yoknapatawpha County was clearly delineated. The white population was neatly divided among patricians and their dependents, who lived in the foothills and valleys; upwardly mobile poor-white tradesmen, who lived in Jefferson and had close ties to hill folk; and poor whites, who lived in the hills. These latter two groups were infiltrated by Snopeses. Byron Snopes, bookkeeper at old Bayard Sartoris's bank, was a "hillman" (*FID,* 7).[31] He was a cousin of Flem Snopes, who recently "to old Bayard's profane astonishment . . . became vice president of the Sartoris bank." These hillmen were "from a small settlement known as Frenchman's Bend"; they were "a seemingly inexhaustible family which for the last ten or twelve years had been moving to town in driblets." Their presence was deplored by the old guard—according to one patrician lady, "General Johnston or General Forrest wouldn't have took a Snopes in his army" (*Sar,* 172, 174). These Snopeses were background figures, but Faulkner's implication was clear: if the Sartorises were dying out, the Snopeses were waiting to take over, and it was going to be a smaller world when they did.

The Sartoris place, screened off from Jefferson by "upland country," occupied the Yoknapatawpha equivalent of Delta land, "a valley of good broad fields." Tradition sat heavy on this house that Colonel John Sartoris had built. "In the nineteenth century," Colonel John had told old

Bayard, "genealogy . . . is poppycock"; that was true "particularly in America," he said, where nobody could be sure of his ancestry. But John Sartoris was proud of his own. People who were not were "only a little less vain" than those who were, and, besides, "a Sartoris can have a little vanity and poppycock if he wants it." Poppycock or not, they seemed locked into their traditions—a "funny family," one friend commented, who were "always going to wars, and always getting killed." They inherited their bizarre behavior from Colonel John, onto whom they projected what his sister, Miss Jenny Du Pre, thought of as "a glamorous fatality." Colonel John's mystique prevented them from dealing with their own lives objectively; the Colonel was an "arrogant shade" who still "dominated the house and the life that went on there" (*Sar*, 6, 92, 167, 380, 113).

John Sartoris was modeled on Colonel William Clark Falkner. Ignoring the Old Colonel's origin as a penniless runaway, Faulkner focused on his image as wartime guerrilla, scourge of carpetbaggers, railroad builder, and duelist. And he made a significant decision. His readers would experience Colonel John as Faulkner had experienced the Old Colonel, who had died eight years before his own birth—through the patchwork of memory and legend passed on by his survivors. The chief source of those legends outside the family was old man Falls, a doddering survivor of the Colonel's regiment. Neither Falls nor anyone else justified Colonel John's successes—as Drusilla Hawk was later to do in *The Unvanquished* (1938)—as paternalistic munificence. The plantation was farmed by tenants and nobody talked *noblesse oblige*. Colonel John was kind to Joby Strother, his slave groom, but Faulkner never approached the Colonel's attitudes on race. The problem went deeper than that and seemed to support a theory Faulkner later enunciated. "Southern men," he said in 1958, had not been able to "bear . . . having lost the war" and had "given up." The effect had been so profound that "even generations later [they] were seeking death"; only the women "could bear it because they had never surrendered."[32]

As Falls regaled Bayard with tales of his father's wartime bravado, Colonel John sounded almost like he had been "seeking death" *before* the surrender, and his postwar efforts to build a railroad and to return the county to white rule were almost as dangerous. Sometimes they also showed how his charisma blinded those who survived him. On election day in 1872 John Sartoris stood at the polling place with the ballot box at his feet and a derringer in his hand; black voters "broke and scattered," and he fired over their heads. Shortly after, shots were heard from the rented room of the carpetbaggers who had organized them, and John Sartoris emerged to apologize to the landlady for "exterminatin' vermin on yo' premises." The tragedies of murder and stolen ballot boxes escaped Falls as he told that story, but he dimly perceived the problem: "when a feller has to start killin' folks, he 'most always has to keep on killin' 'em." That was the tragedy behind Colonel John's heroic image, and it caught up with him not long after. "I'm tired of killing men," he had told Bayard, and, like William Clark Falkner, who had gone to meet Dick Thurmond without a weapon, John Sartoris had been shot down, unarmed, by one of his enemies.[33] From above his grave the Colonel's statue, his face "lifted a little in [a] gesture of haughty pride," now looked "out across the valley where his railroad ran." For the Colonel's descendants the scene was a tricky illusion. With Colonel John dead the valley looked as though his "stubborn dream . . . had shaped itself fine and clear now that the dreamer was [dead and] purged of the grossness of pride." But the pride was still there; that "haughty" gesture the sculptor had caught "repeated itself generation after generation" in the faces of his heirs (*Sar,* 235, 236, 23, 375, 113).

Colonel John's sister, Virginia Du Pre, preserved and embellished the family legends. This formidable lady, modeled on "Auntee" Holland Falkner Wilkins who kept the Young Colonel's house after his wife's death,[34] had come to live with her brother "two years a wife and seven years a widow at thirty." It was Miss Jenny who spun tales of the glamorous, foolhardy Carolina Bayard Sartoris, who had been killed stealing anchovies from a Union quartermaster; as she aged, that "tale . . . grew richer," Faulkner reported, until it became "a gallant and finely tragical focal point" for "the history of the race." To Miss Jenny the tale epitomized the "glamorous fatality" she saw in the Sartoris men; but these men frightened her, too. As she walked through the graveyard, the caption old Bayard had set for Colonel John spoke volumes to her. "Pause here, son of sorrow," it read grandiloquently, "remember death." And Bayard had added a Parthian shot at his father's killer: "By man's ingratitude he died." Later, that Sartoris pride surfaced again when Bayard's grandson John was killed in a hairbrained stunt over enemy territory in France. Ignoring young John's folly, Bayard had inscribed "I bare him on eagle's wings and brought him unto me." No wonder one friend thought Miss Jenny must have been placed on earth "for the sole purpose of cherishing those men to their early and violent ends" (*Sar,* 8, 9, 375, 357).

Coping with these men perhaps accounted for some of her grouchy abrasiveness, which was not always directed at those who could respond in kind. Like Colonel John, Miss Jenny never talked *noblesse oblige,* and sometimes she seemed merciless as she tongue-lashed old Simon Strother, the ex-slave who doubled as groom and butler. "Who was the fool . . .

who thought of putting niggers into the same uniform with white men?" she ranted when Simon's son Caspey came back rebellious from France. Like J. W. T. Falkner, she had crossed class lines to back Vardaman. "Mr. Vardaman knew better," she insisted. Yet Miss Jenny's racism was the bland racism of the aristocrat; her complaint was that she could not *depend* on the Strothers, not that they were brutes, and it masked a vulnerable affection. When Simon was killed at the end of the novel, Miss Jenny sounded like she had lost another member of the family: "Well, that *is* the last one of 'em" (*Sar*, 67, 370).

Miss Jenny's bitterness was directed toward the Sartoris men, whatever their color. The Sartoris man she had had to contend with longest was Colonel John's son, Bayard. Faulkner, ignoring his grandfather's reputation as senator and county boss, had constructed old Bayard's portrait around J. W. T. Falkner's image as banker and man of conservative tastes.[35] Bayard's very prejudices revealed his archaism. Like Colonel John and Miss Jenny, he never justified his life-style as benign paternalism; he merely accepted it, convinced that he had "the triflingest set of folks to make a living for in the whole damn world." Like Miss Jenny, he could be cruel. Simon was a companion to old Bayard as well as a servant, but, when Simon broke the rules of their relationship, Bayard told him he was "a lazy nigger that aint worth his keep." Like Miss Jenny also, he was blinded by his family's myth of a "glamorous fatality." "I have outlived my time," he told his doctor. "I am the first of my name to see sixty years." That was why old Bayard, who despised automobiles, insisted on accompanying his surviving grandson—John's twin who was also named Bayard—on hair-raising rides over country roads. When that wild young man finally wrecked the car and old Bayard's heart stopped, his death wish turned out to be the family's undoing as well as his own: young Bayard, blaming himself, left the state and was finally killed test-flying a faulty plane. For Miss Jenny, that was the final irony: with the last of the Sartoris males gone, old Bayard had been denied "the privilege of being buried by men, who would have invented vainglory for him" (*Sar*, 272, 104, 374).

In Miss Jenny and old Bayard, Faulkner had done a remarkable job of "sublimating the actual" of his family's experience "into apocryphal" and he had accomplished this with convincing objectivity. Now he made a surprising choice: the twins' parents, he reported, were long dead. This decision allowed Faulkner to suggest something that all the Falkner boys must have felt. Murry Falkner had not followed in the very large footsteps of J. W. T. Falkner and the Old Colonel, and Miss Maud came from a different kind of family; "the Big Place," where Auntee Wilkins presided over the Young Colonel's household, had been like a second home to the boys, and that house and its aging inhabitants symbolized the family traditions. Telescoping the Sartoris generations thus allowed Faulkner to write as though he were J. W. T. Falkner's *son* rather than his grandson.

Hard-drinking, humorless young Bayard, last of the Sartoris males, was heir also to the traumas Faulkner thought peculiar to his own generation of southerners. Born to the tragedy that "generations later" had left southern men "seeking death," Bayard had also lived through the war that had ravaged the Lost Generation. Faulkner suggested both traumas through the death of John, whom Bayard thought had been braver than he; Bayard "not only had suffered the . . . injury of having lost a twin," Faulkner commented later, "but also he would have to say to himself, 'The best one of us died, the brave one died.' "[36] Young Bayard's death wish portended the end of the Sartoris line. Sometimes he seemed to accept his existence, throwing himself into the plantation work. He married Narcissa Benbow. Sometimes thoughtless of blacks, he nonetheless seemed the least racist of the Sartorises; Bayard was never seen deprecating blacks, and he enjoyed hunting with Simon's sons, whom he respected as woodsmen. But Bayard never forgot that the braver part of him had died with John; he seemed compelled to demonstrate his courage over and over, usually prompted by alcohol, until with Old Bayard's death he finally created a tragedy he could not accept. Dying in a plane was easier, finally, than facing what had happened to his grandfather.

With young Bayard's death, Miss Jenny outlived all her men, and her attention turned to Narcissa's unborn child. That child would be male and would be named John, she decided. There Narcissa drew the line; he was going to be named Benbow Sartoris, she proclaimed. "Do you think you can change one of 'em with a name?" Miss Jenny asked, but the prospect frightened her. Maybe they were all pawns in some celestial chess game, she wondered, and in a burst of insight suddenly perceived they might be more: that "perhaps Sartoris is the game itself," that it might be "a game outmoded and played with pawns shaped too late and to an old dead pattern." But such insights were too much for Miss Jenny, and she returned to her dream of a *Chevalier sans peur et sans reproche*: of "silver penons downrushing at sunset," of "a dying fall of horns along the road to Roncevaux" (*Sar,* 380).

IV

Sartoris was a remarkable achievement for Faulkner. He had painted the Sartorises lovingly but objectively; he had been able to make his novel "truly evocative" precisely because he had made it "personal." But there were cracks in Faulkner's objectivity in *Sartoris*, and those, too, were there because it was personal. To capture the tone of paternalism, Faulkner had to sample its effect on blacks, and again he turned to his own family's experience. Simon Strother,[37] like Ned Barnett, doubled as groom and butler, and he had the meddlesome disposition that prompted Ned "to take part in everything and boss all of it." Like Ned, Simon was abjectly loyal to his whites, basking in their charisma; like Ned, he was not above conning his white family; and, like Ned, he slid chameleon-like from one role to another, acting by turns the spoiled servant, the irresponsible dependent, and the self-effacing darky.

Simon's preoccupation with the family's "glamorous fatality" paralleled old Bayard's more tragic one. Simon "admired Sartorises," but he "loved horses" and "spoiled" them—just as, perhaps, his white family had spoiled him. He shared Ned's flair for sartorial splendor; Simon looked like the epitome of a comic groom as he sat in the driver's seat of old Bayard's carriage "in a linen duster and an ancient tophat," with the "seemingly . . . incombustible fragment of a cigar at a swaggering angle in his black face." He was a snob about his status. As archaic in his tastes as old Bayard, he shared his patron's preference for carriages over automobiles: "Don't block off no Sartoris ca'iage, black boy," he told one offending chauffeur. That snob persona was just one of many. Waiting for old Bayard at the office, he turned on his "bank" demeanor; with Miss Jenny he switched to "his Miss Jenny attitude" of "gallant and protective deference." Whatever the mask, it covered an arrogance that rivaled John Sartoris's. Simon carried on a running sarcastic dialog with Miss Jenny. Despite his gray hairs he styled himself as a lover, and his lust for exotic Meloney motivated an outrageous scheme to embezzle from his church; when a procession of angry black churchmen confronted old Bayard, Simon lied, "Why Cunnel, you tole me yo'self to tell dem niggers you wuz gwine pay 'em" (*Sar*, 3, 25, 32–33, 271).

Like the Sartorises, Simon stood for a dying life-style, and his adventures with Meloney showed his archaism; that would-be middle-class woman exploited him as casually as a Snopes might. If old Bayard's death was the end of a white way of life, a black way ended when Simon was found in Meloney's cabin with his skull crushed. Faulkner was learning to be more subtle with his blacks. In *Soldiers' Pay* the porter's nostalgia for the South of his youth suggested a one-sided commitment to the Old Order; Simon, who was also sentimental, was self-serving in his loyalty as well. Under the mask of the comic darky, Faulkner now saw a series of other masks, all hiding an agreeable but casually corrupt old man.

When Faulkner turned to blacks of his own generation, however, he had problems. Simon's son, Caspey, wore no masks: he was a shiftless, razor-carrying, crap-shooting epitome of the grossest of stereotypes. The army had affected this black veteran in some of the ways it had affected young Bayard, and Caspey should thus have been an effective foil; he came through only as a clown. "War," he philosophized to Simon, "unloosed de black man's mouf," and he wanted it known he had come home bad. "I don't take nothin' offen no white man no mo'," he said, and argued reasonably, "If us cullud folks is good enough to save France fum de Germans, den us is good enough ter have de same rights de Germans is." But when young Isom asked "how many [Germans] you kilt," it was clear Caspey was suffering from a problem in reality. "I ain't never bothered to count 'um up," he bragged, and told how he and some black friends did away with "about thirty." Now Caspey swelled with pride. "I got my white in France," he confided, "and I'm gwine git it here too" (*Sar,* 62–63, 66).

That speech was a vicious parody of the black American's wartime heroism and postwar aspirations. Blacks like Caspey might or might not have achieved the heroic inefficiency Faulkner ascribed to him, but they had served a country that promised them no rewarding future; and, if Faulkner had relegated him to a labor battalion, Caspey had no need to apologize for his fellows in the 369th Regiment, who were the first regiment of the Allied armies at the Rhine, who served longer than any other American regiment, and who were awarded the *Croix de Guerre* as a unit. Faulkner had understood Simon's decadence, but in Caspey he showed no more ability to understand change than did the Sartorises themselves; what Caspey's portrait called to mind was the propaganda of an earlier war and the avalanche of black characters created by white writers to illustrate the beauties of paternalism. Revolution on the Sartoris place in 1919 did not last long. When Caspey refused an order from old Bayard, the killer of Germans found himself on the ground, while his aging patron stood over him with a piece of stove wood. Caspey seemed convinced, and, like the rest of the nation under Warren G. Harding, he was soon "more or less returned to normalcy" (*Sar,* 199).

For Faulkner that was the point. To young Bayard "normalcy" meant much more than a piece of stove wood; black Caspey lacked Bayard's sensitivity and thus could be re-enveloped in the protective shrouds of paternalism. But what Faulkner had given up to draw the parallel seemed more important, and it was not the only crack in his objectivity. Faulkner continued to display a fondness for racist clichés. "Isom's face," he reported after the boy took a wild ride with Bayard, "was like an open piano"; a young girl's hair was "wool"; the door of a black home opened "emitting a rank odor of negroes"; blacks around the square had "an animal odor"; Simon looked "like the grandfather of all apes" (*Sar,* 78, 344, 341, 119–20, 245). When Faulkner invoked one of those apostrophes to the mule that frontier humorists were fond of, he could not seem to get blacks out of his mind. "Some Homer of the cotton fields," he said, "should sing the saga of the mule and of his place in the South," and, after praising mules for teaching the South "humility, and courage through adversity overcome" and for being "self-sufficient but without vanity," he insisted that the mule was "misunderstood even by that creature . . . whose impulses and mental processes most clearly resemble his." That creature, he said, was "the nigger who drives him" (*Sar,* 278–79).

This bizarre passage identified the central irony of Faulkner's first effort "to recreate between the pages of a book" that world that he "was already preparing to lose and regret." In *Soldiers' Pay* he had not been committed to that task, and his pejoratives, gross as they might be, seemed merely irrelevant; in his new novel he was trying to recreate that world, and the pejoratives were devastating to his credibility. It was as though those old stereotypes of the world of his youth had a life of their own, and, now that he was trying to deal with that world, they surfaced again, blurring the focus of his vision.

Faulkner handled blacks who were less involved with paternalism more effectively. His black yeoman farmers and sharecroppers were stronger and more independent than the Strothers. When one old man found Bayard unconscious in his wrecked car, he advised his grandson, "Don't you tech 'im. White folks be sayin' we done it"; but this sensible young man ignored the risk, pulled Bayard out of the wreck, and took him to town for help. Later, ashamed to go home after old Bayard's death, an exhausted young Bayard stumbled on the cabin of a black farmer on Christmas Eve. Suspicious, the man allowed Bayard to sleep in his barn, but not before he warned him about smoking. "I can't affo'd to git burnt out," he said. "Insu'ance don't reach dis fur fum town." Next morning he refused ten dollars to drive Bayard to the railroad, delaying the departure until he and his wife had passed out the children's "frugal and sorry geegaws and filthy candy" and eaten dinner. "I ain't had no Chris'mus yit, whitefolks," he explained (*Sar,* 207, 342, 347, 346). The tough independence of such men contrasted sharply with the Strothers' spoiled dependence, and they were foils for the Sartorises in another way. Each behaved with honor in trying circumstances; that John Sartoris's great-grandson did not said much about the family's decline.

Still, the real irony of Bayard's encounter with the farmer was not that paternalism had failed the human beings it was supposed to care for; it was that humans like Bayard who had received the most from the system had failed it. As the farmer and his wife shared Bayard's jug "amicably, a little diffidently" on Christmas morning, the true meaning of paternalism in *Sartoris* became clear. Aristocrat and sharecropper seemed to be "touching for a moment" because they were sharing "an illusion — humankind forgetting its lust and cowardice and greed for a day." But that "illusion" could not last. Bayard and the sharecropper's family represented "two opposed concepts, antipathetic by race, blood, nature and environment" (*Sar,* 347). It hardly mattered, on that scale, that greed, lust, and cowardice seemed to be white vices; right or wrong, the world was neatly divided into "antipathetic" camps. Segregation in *Sartoris* was not merely sanctified by society, not merely by law; it was also sanctified by Nature herself. That was where Faulkner's vision defined its limits as he searched for the South of his youth. Faulkner's patricians could be cowardly, they could be greedy — they could even be comic; but all of them were created on the assumption of a heroic potential in the Cavalier life-style, and their thrust was toward a tragic complexity. His blacks were there to illustrate that complexity, and they could do that only by being simple or comic.

Despite such limitations, Faulkner could take pride in his first effort to confront the South of his youth. In most ways he had kept the effort objective, and yet he had managed to make it "truly evocative" because he had had the courage to make it "personal." He had found a South. The problem was the South that he had found: it was no more than the South of his youth. It was an archaic South, whose most recent generation, in the person of young Bayard, was dead before its time, unable to come to grips, as John Sartoris had, with its present; and if the Sartorises were dying out — well, there were Snopeses waiting to take over, and it was going to be a smaller world when they did. Faulkner, unlike Bayard Sartoris, was very much alive, and he was going to have to live in that world. He was going to have to live in the South and live in the twentieth century, and he was going to have to do both at once.

These were the problems that *Sartoris* left unresolved. An aristocrat could cling to an older South, unable to separate its vitality from its tragic archaism; he could abandon that South and make his peace with the South of Snopes; or he could look for yet another South that offered some way to separate the buoyant vitality he still found in its traditions from their moribund weight. There was a long road ahead of Faulkner, and he would visit many Souths before he reached its end; but however far he might travel down that road, he would find no point on his compass that would satisfy him the way the one he had found in *Sartoris* did. The South he had found there was the South of his youth, and however he might struggle to leave it, *it* would never abandon *him.*

4

A Visit with an Idiot

I

Young Bayard Sartoris never had to find his South; he inherited that from the old-style grandfather who raised him. Bayard's problem was finding himself *in* his South, learning how to separate his family's myth of a benign golden age from its Cavalier destructiveness. But Faulkner was painfully aware of one difference between himself and Bayard. Years later, when he was asked why he had not provided the Sartoris twins with a father, he responded that "dramatically . . . the twins' father didn't have a story." This father had come "at a period in history which, in this country, people thought of and think of now as a peaceful one," he explained; ". . . the country was growing, the time of travail and struggle where the hero came into his own had passed." What Bayard's father had to face was that "from '70 on to 1912–1914, nothing happened to Americans to speak of."[1]

That, of course, was too easy, and not simply because of the period that those years framed—the era of Faulkner's youth and of the generation that had produced that era. It was not even that he was ignoring as "peaceful" the traumas that had ravaged that world: the revolt of the rednecks and the petrification of Mississippi into a Jim Crow society, traumas to which his novels since *Sartoris* returned with compulsive precision. Bayard, raised by a grandfather who resembled J. W. T. Falkner, had inherited the South that Faulkner's father, Murry Falkner, and his mother, Maud Butler, had inherited; Faulkner, with his close ties to his own grandfather, had inherited that South, too, but to have dealt with that South in *Sartoris* would have blurred its focus, thrown more distance between Faulkner and that world that he feared he would "lose and regret." But by 1928 J. W. T. Falkner had been dead nearly six years, and Faulkner was increasingly aware of what he had inherited from that generation when "nothing happened to Americans to speak of." The contours of that world could be identified by what happened when some of the situations in *Sartoris* were reversed. What if Bayard had lost grandparents instead of parents? And what if the father and mother who survived them had lost touch with the South they had inherited from those grandparents? Bayard had been tragic because he could not find himself in his South; Faulkner's new hero would have to find a South before he could find himself.

Faulkner's first notion of *The Sound and the Fury* (1929), he later recalled, was that it would "be a story of blood gone bad."[2] The memories of an older world which he drew on for this tale antedated even those he had drawn on in *Sartoris.* When Granny Falkner died in 1906, Miss Maud and Auntee had first thought to keep the children away from the services at the Big Place; it was only at J. W. T. Falkner's insistence that they relented and allowed them a last look at their grandmother.[3] Faulkner's new novel came to him as a vision of "some children being sent away from the house during the grandmother's funeral"; those children "were too young to be told what was going on," but "one of the little girls was big enough to climb a tree to look in the window." That situation made him think of "the blind, self-centeredness of innocence, typified by children," and Faulkner began to see where this fantasy was leading him. He had thought of "how much more I could have got out of the idea . . . if one of those children had been truly innocent, that is, an idiot." From what source, he wondered, would such a child "get the tenderness . . . to shield him in his innocence"? That was how "the character of his sister began to emerge."[4] Faulkner named his idiot Benjy and his little girl Caddy, and he gave them a Snopes-like brother, Jason, who "represented complete evil"; he needed "the protagonist," he decided, "someone to tell the story, so Quentin appeared," too.[5] Benjy and Jason, but not Caddy, would tell portions of the story; Faulkner's protagonist, the oldest of the four, narrated another part (Faulkner later added a fourth) and sounded more like the author[6] — the other Faulkner who had been one more generation removed from his grandparents than Bayard Sartoris.

Immediately he ran into personal problems. In *Sartoris* the practice of "sublimating the actual" of his family's experience "into [the] apocryphal" had served Faulkner well. But Mr. Murry and Miss Maud, unlike the late Young Colonel and Granny Falkner, were very much alive; they were proud, sensitive people, and using them to provide his children with a mother and a father who had lost touch with the South they had inherited from their own parents would be tricky. Faulkner compromised by blurring the reflections of Mr. Murry and Miss Maud (and of other Falkners) in the Compson parents: by magnifying some characteristics, by reversing others, and by weaving in new ones.

Murry Falkner, hardly a failure by most standards, had not lived up to the family's very high expectations; frustrated in his ambition when his father sold the Old Colonel's railroad, he had held a series of respectable but unprofitable jobs, including his then present political sinecure at the University. Jason Compson III, like Mr. Murry, had been ambitious as a youth—enough to earn, as Faulkner's Uncle John had, a law degree; like Mr. Murry, he was sometimes, though not always, remote from his family; and like Mr. Murry (and other Falkners), he was fond of his liquor. But here the similarities shifted in degree. If Mr. Murry had not lived up to family expectations, Mr. Compson, far gone in alcoholism, had sunk below his own. To these qualities Faulkner added some of Miss Maud's. Like Faulkner's mother, Mr. Compson had steeped himself in the ancient writers, and, despite a remoteness Maud Falkner did not share, Mr. Compson supplied what personal warmth his children received.[7]

Caroline Compson had little in common with upright, efficient Maud Falkner except that, like Miss Maud's, her family was less respected than her husband's; it was as though Faulkner had taken his mother's best qualities and supplied Mrs. Compson with their antitheses. "Don't Complain—Don't Explain" read the motto tough-minded Miss Maud had hung in her kitchen, and respect was what she commanded; Caroline got little and complained constantly about it. She was sure "that Father believed his people were better than hers," Quentin remembered, and she thought he was trying "to teach us the same thing";[8] convinced people believed that her Bascomb, not his Compson blood, had produced Benjy, she insisted this child was her "punishment for . . . marrying a man who held himself above me." Quentin was appalled by her sniveling, hypochondriac incompetence. "If I'd just had a mother," he lamented, and, as he tried to put together what had gone wrong, he decided simply that "we were all poisoned" (*SF*, 137, 80, 134, 79).[9]

Such was Faulkner's "story of blood gone bad." Sealed off from the South of old Bayard by the death of their grandparents, the Compson children struggled to find themselves in a South without landmarks. Benjy, three years old when Damuddy Bascomb died, had not developed mentally since that time; what he learned of tradition from her could never be more than a web of unconscious associations. Jason, six when she died (and spoiled by her, Mrs. Compson claimed), could learn of tradition only through his alcoholic father; it was no wonder that he turned against tradition as an adult, joining the world of Snopeses. Caddy, eight when Damuddy died, remembered her grandmother but now had no woman of her blood to turn to but Caroline; the result was a hysteria that drove her from the family. Quentin, ten at Damuddy's death,[10] remembered not only his grandmother Bascomb but also his grandfather Compson; as he struggled to order his life around the values General

Compson had stood for, the tension between those values and what his Bascomb blood stood for was devastating.

Quentin was another of Faulkner's southerners who "generations" after the Civil War still "were seeking death." In a recurring fantasy he saw his grandfather talking with Death itself. Death was "a kind of private and particular friend" of Grandfather's. They were "waiting on a high place beyond cedar trees," and "Grandfather wore his uniform"; beyond them he saw "Colonel Sartoris . . . on a still higher place." The Colonel seemed to have a vision of something lofty and important; he was "looking out across at something and they were waiting for him to get done looking at it and come down" (*SF*, 137). The implications of that fantasy for Quentin were chilling. In it Death was a Confederate officer, someone who knew Grandfather well enough to share Colonel Sartoris's vision. But Quentin could only share that vision by dying, which meant that while he lived meaningful contact with tradition was impossible for him. Cut off from contact with such men, he had been forced into the classic reactionary position: if he meant to live by the traditional values of Grandfather, he had to do it without the one thing essential to *any* tradition, its continuity.[11]

Frustrated, Quentin struggled to re-invoke the tradition through rituals that had long lost their meaning. Cavalier males projected the values they aspired to onto their ladies, then identified their own honor by protecting those ladies. Quentin, unable to project his aspirations onto Caroline, turned to his troubled younger sister; but Caddy, protective of everyone but herself, had wakened to a shattering promiscuity, and now she was pregnant. Quentin, to preserve the family honor, challenged the drummer who was her lover. But the ritual that was supposed to affirm his masculinity affirmed just the opposite: he fainted when he discovered the man's prowess with a pistol. When the drummer disappeared, Caddy married without telling her new husband, Chicago businessman Herbert Head, she was pregnant; the fraud was devastating to Quentin's sense of honor, but Head, who understood nothing about tradition, was worse. "I like you Quentin," he told the boy, because "you dont look like these other hicks"; Head was sure that "there's no future in a hole like this," and he told Quentin patronizingly that he "would like to give you a hand." Still other problems were generated by Quentin himself. Cavalier males, not females, were the legendary promiscuous ones, but it was Quentin, not Caddy, who was chaste. "In the South you are ashamed of being a virgin," he complained, but when Caddy offered to resolve his problem he was paralyzed. "There's a curse on us," he told his sister, and in his emotional state inadvertently defined a universe in which neither they nor Fate, yet somehow both, were responsible: the curse was "not our fault," he said, and in the same breath asked, "is it our fault[?]" (*SF*, 84, 61, 115, 123).

Whoever was at fault, Quentin was afraid to let Fate define its own pattern, and his fear led him into another fantasy. Caddy's promiscuity, which threatened to breed Compsons with drummers and businessmen as well as Bascombs, threatened Quentin's Cavalier values. If social and moral distinctions were lost, what values were left? But what if Caddy's unborn child were Quentin's? That fantasy not only cured Quentin of his virginity, it freed him from the responsibility of accepting Caddy's lover and her husband; and, in a strange way, it also re-affirmed Quentin's own moral values. If Quentin and Caddy could not affirm the values of their ancestors through virtue, maybe they could do it through sin. Quentin wanted them to "have done something so dreadful" everyone "would have fled hell except us" (*SF*, 98). If sin were real, the antithesis of sin had to be real, too, and either eventuality was better than the valueless world Quentin had begun to perceive around him.[12]

Threatening Caddy's lover and quarreling with her husband, Quentin had at least been dealing with facts; now he was lost, and the proof was that he wanted Caddy and his father to accept his fantasy, too. Alcoholic Mr. Compson saw through his lie quickly, but Mr. Compson's cure for Quentin's fears was worse than the illness: Compson wanted to bring his son behind the barrier of nihilism he had erected around himself. "Father was teaching us that all men are just accumulations," Quentin realized, "dolls stuffed with sawdust swept up from the trash heaps where all previous dolls had been thrown away." That was the antithesis of the fixed values of the Old Order, and it brought on a final fantasy. "Father said clocks slay time," Quentin recalled; time was "dead," his father had said, "as long as it is being clicked off by little wheels." Time, for Mr. Compson, was movement away from the vitality of the traditional world of his youth; but Quentin was not prepared to let go of that world. If time for Quentin ceased being clicked off by those little wheels, if it stopped, as it had not stopped for his father, then it could no longer be dead—and Quentin could not become the nihilist his father had become. "The peacefullest words," he told himself, echoing his father's Latin, were "*Non fui*"—and he chanted those words like a priest saying mass. "*Non fui. Sum. Fui. Non sum.*" Dead while he still clove to the values of his grandfather, Quentin would be eternalized as their champion. He was ready for the cedar grove now, where Grandfather in his Confederate grays would be conversing with Death, that "private and particular friend"—the two of them "were always talking," Quentin remembered, "and Grandfather was always right" (*SF*, 137, 66, 136).[13]

The irony of that final fantasy was that it destroyed the values it was created to affirm. Those values demanded that Quentin, the family's eldest son, should assume the responsibilities that his father had abdicated: the responsibilities for the physical and moral welfare of his parents, of Caddy, Jason, Benjy, and of their black dependents. Eighteen years later, on Easter Sunday 1928, Quentin's demoralized family was showing the effects of his failure to assume that responsibility. Caddy, cast off by Head, had disappeared, and Quentin's father was long since dead. Dilsey, the maid who styled herself a black member of the family, struggled to protect Caddy's daughter (named, ironically, for Quentin) and Benjy. Whining Caroline had turned the house over to Snopes-like Jason, who revealed his own moral bankruptcy when, to control Benjy, he had his brother castrated. The paradox of their situation was that this poor man, who at thirty-three still had the mind of a three-year-old, was the only Compson who seemed to remember the values Quentin had died to preserve in himself.

Benjy, who could not reason, lived in a world where any random event might call to his consciousness everything that his memory associated with that event. In Benjy's world the more ordered era of his youth, when his life was programmed by Caddy and Mr. Compson as well as by Dilsey, was as alive as today. Benjy had no faculties for controlling what might or might not cause his memory to take him to that era; and, when such associations were not available, he could not articulate their absence, but he was aware of it. When he heard Caddy's name, he bawled, "slowly, abjectly, without tears." Dilsey's grown son, T. P., drove Benjy to the graveyard every Sunday to leave flowers. As the surrey passed the square "where the Confederate soldier gazed with empty eyes," then turned right toward the graveyard, Benjy was afforded a reassuring view of the town he was raised in; he was gratified to see "post and tree, window and doorway, and signboard," flowing "smoothly . . . from left to right." But on Easter 1928, T. P. was gone; Dilsey's irresponsible grandson, Luster, drove Benjy, and he could find only a broken narcissus for Benjy to carry. "Les show dem niggers how quality does," Luster told Benjy when they reached the Confederate monument, and he turned left, to display his new status as driver downtown, rather than right, toward the graveyard. The result shocked even Luster. "There was more than astonishment in . . . [Benjy's voice], it was horror; shock; agony eyeless, tongueless." His cry escalated "toward . . . [an] unbelievable crescendo" until Jason appeared, turning the surrey toward home. "Dont you know any better than to take him to the left?" he snarled (*SF*, 246, 248, 249).

Faulkner had written *Sartoris* to preserve a world he knew he was going "to lose and regret"; by the end of *The Sound and the Fury* that world was gone. Tradition was a castrated idiot led by a black boy, a broken narcissus the emblem of its doom. To find the South one took a right turn at the eyeless Confederate statue and headed for the graveyard; that pleased the lastborn of the Compson males, even though he could not see Grandfather there in his Confederate grays; along that route things "flowed smoothly, . . . each in its ordered place." But a left turn took one down Main Street, where Snopes was. That was when the "agony" came, "eyeless, tongueless," moving toward that "unbelievable crescendo." To Faulkner in 1929 the world of his youth was a tale bellowed by an idiot. To the Jasons that tale was "just sound" and signified nothing, and Damuddy and Grandfather were not there to hear it. Still, "it might have been all time and injustice and sorrow become vocal for an instant by a conjunction of planets" (*SF*, 249, 224).

II

As Faulkner reminisced in the 1950s about the story of the children who had been shut out from their grandmother's funeral, he seemed vaguely unsatisfied. He had written the tale from Benjy's point of view first, he said, and when he had finished he was sure it was "incomprehensible"—he had to try it again. "I decided to let Quentin tell his version," he said; but when he had finished he realized "there had to be the counterpoint, which was the other brother, Jason," and when Jason was finished the tale somehow seemed less effective than ever. "By that time it was completely confusing," he decided, and there was only one choice left: "I had to write another section from the outside with an outsider, which was the writer, to tell what had happened." Thus, "by making myself the spokesman," Faulkner said, he had hoped "to gather the pieces together and fill in the gaps."[14] But as Faulkner set out to make himself the spokesman, his mind was on Dilsey Gibson, the Compsons' mammy who, like Caroline Barr, styled herself a black member of the family. In *Soldiers' Pay* he had used the primitive faith of old-style blacks like Dilsey to balance the decadence of his moderns; a similar gallery of blacks might be far more effective in this new novel, which, unlike *Soldiers' Pay,* dealt directly with the class of whites traditionally charged with the responsibility for them. What was more, Faulkner had been learning a great deal about blacks since *Soldiers' Pay,* and he was anxious to put that knowledge to use.

Jason, who "represented complete evil" to Faulkner, was a walking dictionary of pejorative clichés. No one could accuse Jason of singling blacks out for special attention; they had to take their place in a spectrum that included Jews, women, cotton brokers, musicians, and outsiders of all types. Jason had a theory of black behavior: "When people act like niggers," he declared, "no matter who they are the only thing to do is treat them like a nigger" (*SF*, 141). Jason's prejudices, in short, were red-neck prejudices, and Faulkner gave them to the "evil" Compson brother to show how Quentin's attitudes were more tolerant. Frustrated in his efforts to learn about his heritage through his white kin, Quentin turned pathetically to his memories of Dilsey and her husband Roskus, and his attitudes were an index of Faulkner's growing awareness of black character. "They come into white people's lives . . . in sudden sharp black trickles," he decided, "that isolate white facts for an instant in unarguable truth like under a microscope" (*SF*, 133).

Quentin's memories of Dilsey and Roskus isolated the best and the worst in Quentin. After buying the flatirons with which he would drown himself, Quentin boarded a Boston trolley and found the only empty seat was beside a black man; seating himself, he reflected on the ironies of being a southerner at Harvard. "I used to think that a Southerner had to be always conscious of niggers," he mused. "I thought that Northerners would expect him to." That had made Quentin *self*-conscious. "When I first came East I kept thinking You've got to remember to think of them as coloured people not niggers." Fortunately, he recalled, he "wasn't thrown with many of them"; otherwise, he would have "wasted a lot of time and trouble" before learning "that the best way to take all people, black or white, is to take them for what they think they are, then leave them alone." That was how he had come to understand that what he had thought of as "nigger" attitudes were surface attitudes. "I realized that a nigger is not a person so much as a form of behavior; a sort of obverse reflection of the white people he lives among" (*SF*, 67).

Quentin's attitude was the antithesis of that of Jason, who thought people who acted like "niggers" must *be* "niggers": if the "nigger" was on the outside, then the man inside was what he thought himself to be, not what others claimed. Quentin had thus taken the first step toward seeing behind the black mask; but these musings suggested an awareness that was beyond his power to sustain. When Quentin found himself suddenly homesick for the Gibsons, he launched into a dissertation on black character that showed that he had not learned how to separate his affection for individual blacks from the sentimental patrician postures of his youth.

For Quentin black character was an anomaly, a strange "blending of childlike and ready incompetence and paradoxical reliability." It "tends and protects them it loves out of all reason," he decided, and yet it "robs them steadily and evades responsibility and obligations," often "by means too barefaced to be called subterfuge even." Blacks "taken in theft or evasion" sometimes responded like aristocrats—"with only that frank and spontaneous admiration for the victor which a gentleman feels for anyone who beats him in a fair contest." Like aristocrats they were above prejudice; sometimes they even seemed paternalistic toward the whites who were supposed to take care of them, displaying "a fond and unflagging tolerance for whitefolks' vagaries like that of a grandparent for unpredictable and troublesome children" (*SF*, 68). Thus, although Quentin was able to see the paradox, even the role reversals in some black behavior, his notions of black character were sentimental and traditional, and the mask was still there. The blacks he was fond of were lazy and affable, tolerant and licentious, and he could not see that they were the "obverse reflection" of the Old Order's self-image of protectiveness and responsibility.

Quentin's reflections anticipated the image of black character that emerged as Faulkner set out to make himself the spokesman for the Compsons' tragedy. Hardworking Dilsey showed the "paradoxical reliability" that "tends and protects them it loves out of all reason"—nobody without it could have managed to live with the Compsons so long!—and Dilsey, her mind on the upcoming Easter service, blended these qualities with something like the "childlike and ready incompetence" of which Quentin spoke. Aging and overworked, she had been looking forward to this Easter, and when she saw that the morning sky was cloudy she reacted with "a child's astonished disappointment." She was no model of household efficiency; she forgot Caroline's hot water bottle, and she forgot she had to get Luster to work taking care of Benjy. She was evasive, too, making excuses for Luster's disappearance and lying to Caroline that "de water jes dis minute got hot." But Dilsey would not have been there at all if she had not had her share of "fond and unflagging tolerance for whitefolks' vagaries." As Faulkner unfolded Dilsey's character, he found another archaism that he had not foreshadowed through Quentin's musings. Fat, hulking Dilsey did not resemble tiny Caroline Barr physically, but she identified herself as completely with the Old Order as Mammy Callie did. Like Mammy, Dilsey never doubted that she and her black family were members of that white family. "Dese is funny folks," Luster said of the Compsons, and added, "Glad I aint none of em." But Dilsey snapped, "Aint none of who? . . . Lemme tell you somethin, nigger boy, you got jes as much Compson devilment in you es any of em." Dilsey had a privileged servant's contempt for "trash white folks," and she talked about people of her own race as a white would. To Dilsey, the younger generation of blacks were just "dese here triflin young niggers" (*SF*, 207, 210, 214, 226, 225).

That side of Dilsey sounded traditional enough; but Faulkner intended to make more than another literary darky out of the Compsons' servant. Quentin found himself puzzled by still another side of black character, and as he struggled to understand it he might have been talking about Sherwood Anderson's blacks in *Dark Laughter.* Blacks sometimes came through to Quentin as "just voices that laugh when you see nothing to laugh at, tears when [there is] no reason for tears." They were impossible to measure against white standards. Blacks he had known would bet on anything, even "the odd or even number of mourners at a funeral." And he recalled a strange incident in which "a brothel full of them in Memphis went into a religious trance [and] ran naked into the street"—it had taken "three policemen," he remembered, "to subdue one of them" (*SF,* 133). Traditional blacks were the "obverse reflection" of the image of the domineering Cavalier; blacks like the exotic primitives of *Soldiers' Pay* seemed to be the "obverse reflection" of Puritan poor whites.

It was here that Faulkner's narrative focused itself as he struggled to place the Compsons' decadence in perspective. The blacks who surrounded Dilsey were as musical as Carl Van Vechten's jazz men. Luster was fascinated by a visiting white man who played a hand saw; Dilsey, anticipating the Easter service, sang "something without particular tune or words, repetitive, mournful and plaintive, austere"; at the black Easter service she sang hymns with the congregation and joined the incanted responses to the minister's oration. That congregation dressed itself for Easter in gaudy finery. As Dilsey walked toward the church, the older men she met were somber "in staid, hard brown or black," but their formality was broken by "gold watch chains and now and then a stick"; the younger men wore "cheap violent blues or stripes and swaggering hats"; Dilsey's daughter Frony wore "a dress of bright blue silk and a flowered hat." Dilsey herself, for this Easter at least, had, like Ned Barnett, let clothes become her foible. She had emerged from her cabin Easter morning three times in different costumes, her final one including "a pair of soiled white elbow-length gloves" and "a maroon velvet cape with a border of mangy and anonymous fur above a dress of purple silk." But Faulkner's most obvious link with the "primitive" tradition was the "weathered church" itself, which looked like it could have been painted by Joseph Pickett or Edward Hicks; it "lifted its crazy steeple like a painted church, and the whole scene was as flat and without perspective as a painted cardboard set upon the ultimate edge of the flat earth, against the windy sunlight of space and April and a midmorning filled with bells" (*SF,* 210, 226, 225, 223, 207, 227).[15]

Faulkner's black exotics in *The Sound and the Fury* might seem comic and bizarre; but the church showed the true meaning of exoticism for Faulkner in this novel. In *Soldiers' Pay* Faulkner had kept his primitives' spirituality at a distance—it was something remote, beyond a white person's comprehension; now he was ready to show what that spirituality meant, and, as his blacks went through their Easter rituals, he revealed a remarkable sensitivity to what they were doing.

Sixteen years before *The Sound and the Fury* was published, James Weldon Johnson set down some of his impressions of the "big meetings" he had attended in his novel, *The Autobiography of an Ex-Coloured Man*. "As far as subject matter is concerned," Johnson's narrator recalled, "all of the sermons were alike." For example, they were ambitious: "each began with the fall of man, ran through various trials and tribulations of the Hebrew children, on to the redemption by Christ, and ended with a fervid picture of the judgment-day and the fate of the damned." He was particularly impressed by one master orator, who had a "magnetism and an imagination so free and daring that he was able to carry through what the other preachers would not attempt." This man "knew all the arts and tricks of oratory, the modulation of the voice almost to a whisper, the pause for effect, the rise through light, rapid-fire sentences to the terriffic, thundering outburst of an electrifying climax," and he had "the intuition of a born theatrical manager."[16]

Such a man was Faulkner's Reverend Shegog, who presided over Dilsey's Easter service. This St. Louis evangelist was no untutored child of nature. Like Johnson's orator, he represented a generation of sophisticated technicians aware of the rhetorical as well as the spiritual aspects of his calling; but Shegog's power went further. His expertise was in invoking the "primitive" response from his audience. Black sermons often began with learned dissertations in textbook English, as though preacher and congregation felt compelled to force themselves through layers of white man's religion before confronting their own.[17] Faulkner's minister had made that a part of his technique. His appearance was unprepossessing—he was "undersized"; when the congregation saw him, they were shocked, and when he spoke their disappointment increased. "He sounded like a white man. His voice was level and cold." Reverend Shegog won them over somewhat with his rhetoric, with "the virtuosity with which he ran and poised and swooped upon the cold inflectionless wire of his voice." Still, it was a white sermon, and when Shegog paused at last they were sure Easter was over for them this year. It had hardly begun. Shegog's voice echoed through the church: "Brethren." They were electrified. Now, moving back and forth across the pulpit, he spoke in a voice that carried "a sad timbrous quality like an alto horn, sinking into their hearts and speaking there again when it had ceased in fading and cumulate echoes." Later they would not remember "when his intonation, his pronunciation, became negroid"; they would only remember how that voice "consumed him, until . . . there was not even a voice but instead their hearts were speaking to one another in chanting measures beyond the need for words" (*SF*, 228, 229, 230). Faulkner's black Easter had become primitive.

Faulkner had said little about the content of the "white" part of Shegog's sermon; he spelled out the "negroid" part in detail.[18] "I got de ricklickshun en de blood of de Lamb!" Shegog shouted; "his whole attitude" was now "that of a serene, tortured crucifix." What followed sounded like a confused tirade composed of fragments of biblical stories. "I hears de weepin en de cryin en de turnt-away face of God: dey done kilt Jesus. . . . I sees de doom crack . . . en de arisen dead!" But those fragments electrified his audience as no white oration could have. "I sees, O Jesus!" someone cried. "Oh I sees!" That was the climactic moment in Faulkner's tragedy of the children who had been shut off from their family's past. Now it was clear where the source of the love and responsibility in that family was, and the irony suddenly emerged that it was the children's mammy who was the heir of what they had lost in Damuddy and Grandfather. As Shegog's voice ceased, Dilsey emerged as the novel's moral center. She sat there "bolt upright," Faulkner reported, "crying rigidly and quietly in the annealment and the blood of the remembered Lamb" (*SF*, 229–31).

Faulkner had prepared Dilsey carefully for that moment. As Dilsey waited on whining Caroline, as she struggled to protect Caddy's daughter and Benjy from Jason, it became clear who was holding the family together; those efforts, for which she received little in return, had come to be her whole life. Frony, watching her mother's distracted conduct that morning, was aware of what Shegog's sermon would do for her. "He'll give her de comfort en de unburdenin." Now Dilsey had that peace. What she had known only in her mind before Shegog's sermon had now penetrated her heart: God might be crucified, but He was also risen, and every sinner was redeemed. Now Dilsey could talk about the tragedy she had spent her life witnessing. "I've seed de first en de last," she was now ready to admit, and she could accept that vision because Shegog had shown her what he had seen: "de power en de glory" (*SF*, 227, 231). Such, at last, was the meaning of the soaring black spirituality Faulkner had discovered in *Soldiers' Pay*. In that novel blacks had taken "the white man's . . . remote God and made a personal savior of Him," but the quality of their experience had remained a mystery. Dilsey and Shegog showed how blacks found God. Whites approached God as Shegog had in the first part of his sermon, through a purely cognitive effort; that, it appeared, might be what was wrong. But blacks had memories of an earlier culture, and they could experience God in the depths of their beings.[19]

With her response to Shegog's sermon, Dilsey became more than just another black portrait: her spirituality threw the decadence of Faulkner's aristocrats into stark relief. Faulkner enlisted the Old Order's very virtues to perform the task. That the Compsons' mammy had retained her sense of place in a crumbling order while her whites had lost theirs was ironic; that her faith had inspired her to protect those who styled themselves as her protectors intensified that irony painfully.[20] That Quentin's mammy succeeded where that posturing, compulsive young man failed identified the collapse of those values; and that whites like Jason and Caroline rejected her humanity was the hallmark of that collapse. By fusing in Dilsey the hackneyed conventions of the plantation darky and the exotic primitive, Faulkner had produced a portrait that was far more than the sum of its parts.

He had, in fact, succeeded so well that it was easy to overlook the points where his vision had failed. Quentin's ability to see blacks as "the obverse reflection" of the whites around him showed that Faulkner was growing more subtle, but Faulkner's subtlety seemed to stop at the point that Quentin's did. The Homer of the cottonfields, who in *Sartoris* had spoken of similarities between blacks and mules, still lurked in the wings. Faulkner seemed compelled to talk about how he thought blacks smelled and how they reminded him of animals. A "rich and unmistakable smell of negroes," he wrote, pervaded the black section of Jefferson; and if the Reverend Shegog's sermon was "primitive," that seemed to have something to do with proximity to the jungle. Shegog had "a wizened black face like a small, aged monkey," Faulkner wrote; it sat atop a "monkey body," and Shegog's black audience, repelled at first by his appearance themselves, listened to this seer "as they would have to a monkey talking" (*SF*, 226, 228).

Other notes were just as discordant. It was one thing to suggest Shegog was rejecting "civilized" sophisms for "primitive" faith when his voice "became negroid"; but in *The Sound and the Fury* the signal that Shegog was becoming "negroid" came when he began to talk in dialect. The use of dialect was not the problem; Sterling A. Brown used simulated phonetics with telling effect to capture the tenor of black folkways in *Southern Road* (1932), and folklorists like Zora Neale Hurston used similar techniques years later to record sermons and spirituals. The problem was the way Faulkner used dialect. Sometimes he seemed to be capturing the very note and trick of these patterns;[21] at other times his intention seemed merely comic and his phonetics overdone, recalling the tirades of Mammy Callie Nelson in *Soldiers' Pay*. "Listen, breddren!" he had Shegog saying. "I sees de day. Ma'y [Mary] settin in de do' wid Jesus on her lap. . . . Like dem chillen dar, de little Jesus. . . . I . . . sees de sojer face: We gwine to kill! . . . I hears de weepen en de lamentation of de po mammy widout de salvation en de word of God!" Faulkner had given Shegog a moving sermon; but those *de's* and *dem dar's* operated on another level, calling attention to themselves, instead of to Shegog's words. Dilsey's speeches revealed similar problems. Before the service Dilsey reflected solemnly of the black congregation that "what dey needs is a man kin put de fear of God into dese here trifling young niggers." When Luster was taking Benjy to the graveyard, she threatened her grandson that "ef you hurts Benjy, nigger boy, I dont know what I do. You bound fer de chain gang." In a favorite passage of Faulkner's,[22] she reflected on her fate: "I does de bes I kin. . . . Lawd knows dat." Even when Dilsey was weeping over her vision of the Compsons' tragedy, she sounded like Bones talking to Mr. Interlocuter: "I've seed de first en de last" (*SF*, 230, 225, 248, 247, 231).

What Faulkner was attempting sounded no different from what plantation propagandists had been doing for years: buffoonery made it easier for some whites to take black characters seriously, and at the same time it relegated them to the level of dependents who needed protection. Faulkner, as usual, had something else in mind, too. Shegog and Dilsey were talking about tragedy, but the words they used were the words that had been employed to evoke patronizing laughter in whites for more than a hundred years; the result was a tension very much like the tension resulting from what has been called "black humor," a bizarre mixture of humor and pathos. In one sense that tension served Faulkner's purposes, illustrating his central irony: if Dilsey and her preacher (when he was using "negroid" speech) seemed to sound ignorant, it was all the more ironic that they could find tradition when the Compsons could not. But that irony applied only to the white Compsons, and, although it effectively revealed their tragedy, Shegog's and Dilsey's dialect created another unintended irony that undermined Faulkner's purposes. Readers who laughed at Dilsey were patronizing her, and that suggested the central problem Faulkner was having with Dilsey. Faulkner had shown, through his black Easter, a side of Dilsey that the Compsons never saw; but there was a side of Dilsey that Faulkner did not seem to see either, and its existence revealed with precision the limits of his vision in *The Sound and the Fury*.

The experiences of Dilsey's black family had been, in their own way, as tragic as those of the white family she thought herself a member of; but Faulkner seemed almost indifferent to Roskus, Dilsey's husband, or Frony, her daughter. Superstitious, arthritic Roskus seldom appeared. Faulkner suggested in Benjy's narrative that he had died, but that was about all he had to say on the subject, and Dilsey, her mind on the tragedy of her relationship with the Compsons, never mentioned her dead husband once on Easter 1928. Faulkner showed even less interest in Dilsey's daughter. Frony seemed to have been abandoned by Luster's father; but that did not seem to be on Dilsey's mind this Easter, either.[23] Both of those tragedies were more personal than what Dilsey had suffered with the Compsons, and both suggested compelling reasons for her resignation as she ground out her meager living with them—even some reason, perhaps, for having adopted this white family. Because Faulkner ignored those motives, he left Dilsey's meaning for the Old Order, finally, obscure.

Dilsey, after all, was an anachronism. By 1928 the number of blacks who thought of themselves as black members of white households was shrinking rapidly. The heroism Faulkner had programmed for Dilsey, moreover, was nebulous enough. Such heroism meant accepting a system that relegated blacks to subordinate positions at birth, and Dilsey accepted her position even when the white Compsons abandoned their own responsibilities to the system. Shegog's mysticism, which allowed Dilsey "de comfort en de unburdenin," was similar. Such religion might help Dilsey find the faith to stay with the Compsons, but it did not encourage her to pursue a better life. And what was true of Dilsey was true of Faulkner's entire black community in *The Sound and the Fury*. The Easter congregation reached the heights of mystic transcendence; but no young black emerged to give that transcendence social meaning. What heroism the novel showed was the passive heroism of endurance. The virtues Faulkner assigned his blacks were ambiguous, implying the necessity of the caste system whose epitaph he seemed to be writing.[24] For Faulkner in 1929 the word was still segregated.

That was the final irony of *The Sound and the Fury*. It had taken courage for Faulkner to write about the failures of his parents' generation; but when he turned to Dilsey to help him "gather the pieces together" he had gone too far; he had made his novel's climactic moments depend on his ability to portray blacks whose spirituality balanced the Compsons' nightmarish tragedy, and here his vision was limited. The Compsons provided a frightening image of the attenuation of the old values, but Faulkner's black community provided an ambiguous and patronizing image of the available alternative to those values.

No wonder that by the end of *The Sound and the Fury* tradition was an idiot, bellowing everything and nothing. Faulkner was not aware of it, but his new novel raised questions that he would have to struggle with for a long time, questions that travestied both the pathos of the Compsons' collapse and the mellow nostalgia with which he had portrayed the decline of the Sartorises. If the Sartoris's Simon Strother chose after manumission to retain his place in a declining life-style and if Roskus and

Dilsey chose to stick it out with the Compsons long after that life-style had lost its moral center, their choices might accurately reflect the choices made by some blacks; but the fashion in which Faulkner presented those characters obscured the significance of their choices. Were slavery and servitude really as rewarding as Faulkner's black portraits made them appear? With the protective Sartorises and Compsons gone, what were blacks going to do when they had to deal with Snopeses? And what about blacks who had not been fortunate enough to have white families? In *Sartoris* and *The Sound and the Fury* Faulkner had provided terrifying accounts of the decline and disappearance of the values of the Old Order; but those novels revealed, also, that he had yet to face the consequences of those values for the human beings over whose lives they presided. He was still searching for his South.

5

A Visit with Some Puritans

I

By 1929 William Faulkner had twice visited the South of his youth. He had visited the Sartorises, whose bland decadence travestied their family legends of an older, more vital South; and he had visited the Compsons, whose genuine decadence was like a nightmare of what a southern aristocratic family might become. But there remained a South he had not visited. If the Sartorises and Compsons were on their way out, it was important to know what kind of South was going to emerge without them. To most Mississippi aristocrats in the first half of the century, the answer to that question was simple: a South without aristocrats would be a catastrophe. Blacks, aristocrat William Alexander Percy thought, were "a race which at its present stage of development is inferior in character and intellect to our own"; without the leaven of paternalism people of Percy's class believed that blacks would sink back into a fatal irresponsibility that had nearly destroyed them during Reconstruction—and Percy was sure they would be brutalized without the aristocrats as a buffer against "the lawlessness and hoodlumism of our uneducated whites."[1] Faulkner's first visit to this South without aristocrats had been postponed when he set aside *Father Abraham*, his tale of Flem Snopes; in the novels that followed *The Sound and the Fury* he seemed to be avoiding it; few blacks caught his attention as he explored the isolated redneck country of Frenchman's Bend in *As I Lay Dying* (1930), or as he returned for a round with decadent aristocrats in *Sanctuary* (1931). But a visit to that country, Faulkner knew, was important at this stage of his career; it would show him a great deal about tradition and what it meant to lose it.

Thirty years later Ralph Ellison attempted to express his own vision of the causes of white racial prejudice. Its roots, he thought, were in white American religious experiences. White Americans, Ellison thought, suffered a "Manichaean fascination with the symbolism of blackness and whiteness"; blacks had been "caught up associatively in the negative side of this basic dualism of the white folk mind, and . . . shackled to almost everything it would repress from conscience or consciousness." For these reasons it was "almost impossible for many whites to consider questions of sex, women, economic opportunity, the national identity, historic change, [or] social justice . . . without summoning malignant images of black men into consciousness."[2]

The "white folk mind" Faulkner had been dealing with so far was a Puritan mind. To most observers, a Manichaean theology[3] that made Satan the equal in power to Jehovah seemed alien to Mississippi's Puritan folkways. But like other aristocrats, Faulkner looked on Puritans with distrust; in 1957 he was still complaining about them. Southern Baptists were "bucolic" and "provincial," and it was a "debatable question whether that sort of Baptist believes in God or not." Their faith "came from times of hardship in the South," he said, "when there was little or no food for the human spirit," and he was ready to admit that it represented "the human spirit aspiring toward something"; the trouble was that "it got warped and twisted in the process" and had become "an emotional condition that has nothing to do with God or politics or anything else."[4]

In 1930 Faulkner seemed to think this faith had something to do with those poor whites who were the political enemies of Mississippi's aristocrats. The God that the citizens of Frenchman's Bend worshiped in *As I Lay Dying* certainly sounded Manichaean. Farm wife Addie Bundren lay in her bed listening to "the dark land talking." It spoke to her "of God's love and His beauty"; but to Addie, God had also created the darkness she lay in, and the dark land also spoke to her of "His sin." Instead of elevating Satan to a status equal with God, as Manichaeans traditionally did, Addie had elevated darkness to equality in God's sight with the light. Sinning for this maverick Puritan was an act as sanctified, and as necessary, as loving; the real sin was living in a world without darkness. That was why her husband Anse "was dead" in her eyes; no one was really alive who did not recognize this duality of God's nature. Addie's duty to herself was to stay alive, which meant accepting both sides of her nature, and that was the reason for her love affair with preacher Whitfield. Loving Whitfield meant exploring the darkness: Whitfield "was the instrument ordained by God who created the sin, to sanctify that sin He had created" (*AILD*, 116–18).

Christians long ago found ways to resolve the dilemma with which Addie struggled. "In the beginning," St. John wrote, "was the Word, and the word was with God, and the Word was God." The Word was greater than Addie's "terrible blood" and comprehended it. He was greater than "life" or "light," and was their source: "In him was life; and the life was the light of men." But God knew how impossible it would be for men to accept this light without the Word. If light was to shine in the world, it had to shine "in the darkness," even though "the darkness apprehended it not." That was the meaning of Christ's sacrifice, why "the Word became flesh, and dwelt among us": Christ allowed humans to free themselves of final responsibility for either light or darkness by projecting both onto Him. Because that responsibility was His, humans like Addie could remain as God had made them, creatures of evil as well as good. That was why Manichaeanism became heresy in His church.

Faulkner had learned about his rednecks in *As I Lay Dying.* They were Puritans, but they were a special brand of Puritan, reliving an old heresy. They acknowledged the ascendency of God's light, but they were fascinated with the darkness. The darkness made them real and, in a strange way, also made them Puritans. It was this poor-white preoccupation with darkness that explained to Faulkner what a South without aristocrats would be like when the two plebeian classes clashed. Dilsey and the black singers of *Soldiers' Pay* had found the Word, but whites like Addie and Anse were lost. The Word was dead for them, and they were fated to struggle the rest of their lives with a dilemma they could never resolve.

In *Sanctuary* Faulkner added another dimension to these speculations with the brief tale of "a Negro murderer" whose crime resembled that of Nelse Patton in Oxford twenty-three years earlier. Addie, who had killed no one, was shut off from the Word; the murderer in *Sanctuary* was trying to find it, but he, at least, thought he had been shut off from it by people like Addie. This man had killed his own wife, not a white woman; but her fate was as bizarre, and as exotic, as Mattie McMillan's had been at Patton's hands. He had "slashed her throat with a razor so that, her whole head tossing further and further backward from the bloody regurgitation of her bubbling throat, she ran out of the cabin door and for six or seven steps up the quiet moonlit lane." Now, as people passed the jail, they heard him as he struggled for the Word, singing in a "rich, sourceless voice" that echoed from the dark above them, and from time to time they heard his complaint: "One day mo! . . . Say, Aint no place fer you in heavum! Say, Aint no place fer you in hell! Say, Aint no place fer you in jail!" (*S*, 91–92).[5] If he had not found the Word, this black murderer's hymn-singing suggested that he, like the Easter congregation in *The Sound and the Fury*, knew where to find it; but to the Manichaean whites who passed the jail he was the incarnation of darkness. He had been "caught up associatively," as Ellison would later explain, "in the negative side of [a] basic dualism of the white folk mind, and shackled to almost everything it would repress." That was why there was no place for a black murderer in heaven, in hell, or even in jail. And that was the p lem with which Will Mayes, the black night watchman who was lync in "Dry September" (1931),[6] had had to cope.

The whites who wanted Mayes lynched were Puritans who suffered from repressed guilt and repressed sexuality. John McClendon, who led the lynching party, beat his passive wife when she waited up nights for him to come home. Minnie Cooper, who claimed Mayes raped her, was a spinster of thirty-nine.[7] White Puritan men like McClendon cast themselves as male defenders of light; they projected what they wished to "repress" onto blacks and women and punished it vicariously. White Puritan women like Minnie Cooper faced emotional starvation because Puritan men denied them sexuality; such women projected that pent-up sexuality onto blacks like Mayes and borrowed it back. In a way such people were more tragic than Mayes. The ritual of scapegoating, applied to this man instead of Christ, afforded them no relief from the pervading spiritual drought. Mayes faced death bravely, but these whites were condemned to live—and to live without the Word.

Writing about the redneck South was helpful to Faulkner. He felt no commitment to people like McClendon and Minnie Cooper, and he could see why they hated blacks and feared them. But the work that showed where these observations were taking him was about Quentin Compson and his belief that "a nigger is not a person so much as a form of behaviour; a sort of obverse reflection of the white people he lives among." The behavior of Nancy, Faulkner's black heroine in "That Evening Sun" (1931),[8] looked like the obverse of McClendon's and Minnie Cooper's behavior; Nancy had found the Word and rejected it. Superficially, this story looked like no more than another round with the darky and the exotic primitive. Washwomen like Nancy, Quentin remembered, were a colorful class; they wore turbans and balanced bundles of clothes on their heads. When Nancy went home to "Negro Hollow," she had to cross a deep ditch, climb out, then struggle through a gap in a fence; he was amazed that her bundle "never bobbed or wavered" (*CS*, 290). Nancy's exoticism did not end with the wash bale; a jailor thought she took cocaine. And Nancy was the only sexually passionate black woman about whom Faulkner had written; if her husband were unfaithful, she said, "I'd cut his head off and I'd slit her belly." Her husband felt the same way. Nancy was involved with Stovall, a Baptist deacon who made her his whore and got her pregnant. When she found a hog's bone on her table, she was sure it was a sign that her husband meant to kill her. Nancy's fears became so strong that she moved into a trancelike state; she felt no pain when she absently placed her hand on the globe of a hot lamp. "She was looking at us," Quentin reported on another occasion, "only it was like she had emptied her eyes, like she had quit using them" (*CS*, 295, 302). But like the exotics of *Soldiers' Pay* and *The Sound and the Fury*, primitive Nancy was a believing Christian, and her faith showed what happened when blacks were "caught up associatively" in the "basic dualism of the white folk mind."

Her will broken, Nancy was ready to believe a white man's verdict on her behavior, and she adopted the attitude that Stovall implied by the way he treated her: that a "nigger" is subhuman and hence not morally responsible. "I just a nigger," she repeated compulsively. "It aint no fault of mine" (*CS*, 309). But Nancy was more than that, and, as the Compson children struggled to understand, Faulkner used their ingenuous comments to suggest what was happening. Five-year-old Jason was preoccupied with what it meant to be a "nigger." "I aint a nigger," he announced; then, as though he doubted that, he asked, "Are you a nigger, Nancy?" (*CS*, 298). That was the most important question about Nancy, and she had more than one reason for answering it, because the husband she loved and feared was not ready for her to accept that status. Appropriately enough, the name of that razor-wielding husband was Jesus.

The children saw this "short black man, with a razor scar down his face," only once, at the beginning of the narrative; but his presence pervaded Faulkner's story. Faulkner used that Spanish name to suggest how Jefferson was polarized into blacks who could receive the Word and whites who could not. Quentin remembered that his father "told us to not have anything to do with Jesus" and that "father told Jesus to stay away from the house." Banished from this white home, Jesus soon left Nancy, too. "Done gone to Memphis," she said. Predictably, he was a fugitive: "Dodging them city *po*-lice for a while, I reckon." When he came back to Jefferson, the Compsons, who had failed in their responsibility to Nancy, were among the last to know. "He's not here," Mr. Compson insisted when Nancy told him Jesus was waiting in the ditch near her cabin. "There's not a soul in sight." Blacks were the first to know. "Some Negro," Mr. Compson told his wife, "sent her word that he was back in town." If Jesus was going to exist in Jefferson, Jason was right: "Jesus is a nigger." Why not "a short black man, with a razor scar down his face"? (*CS*, 290, 292, 293, 306–7, 297).

For blacks like Nancy Jesus was never far away. One night her moaning woke the children—she was calling, aloud, for Jesus. It was "like this," Quentin said: "Jeeeeeeeeeeeeeeeesus"—and suddenly he understood: "It's the other Jesus she means." Nancy, who knew in her heart that there was no distinction between her guilt against her husband and her guilt against God, had confused one Jesus with the other, fusing them into an unbearable image of divine vengeance. Her confusion explained why blacks could experience the Word when whites could not. White Christians polarized their experiences of good and evil into a Christ who was god of light and a Satan who was god of darkness, a Christ of the spirit and a Satan of the flesh. Blacks like Nancy (and maverick whites like Addie Bundren) fused that experience into a Christ who stood for both. Nancy's Christ was a jealous husband who behaved like the devil-nigger of the "white folk mind": a god of love who was also a god of vengeance, a god of light who was also a god of darkness. That was how blacks could find the Word when whites could not. The light, St. John said, "shineth in darkness"; "Jesus," Jason perceived, "is a nigger." For Nancy, the Word had literally been made flesh.[9]

"That Evening Sun" showed how Faulkner was developing. Writing about the South of his youth, he had not understood the bland racism behind the Sartorises' tolerance of the Strothers or the Compsons' dependence on Dilsey. Substitute-maid Nancy and night-watchman Will Mayes were not a part of that South, nor was Stovall nor Minnie Cooper nor John McClendon. Faulkner could see them more clearly, and he had been able to evolve a subtle myth that explained the psychological interplay between Puritan rituals of scapegoating and blacks who were caught up in them. White Puritans admitted only the enlightened side of their natures; they banished their darker natures by projecting their guilt onto blacks, women, and children, then punishing that guilt vicariously. Blacks caught up in this white ritual might do a similar thing: they might try to avoid their own guilt, as Nancy tried to do, by accepting that dark, subhuman identity and punishing themselves. But blacks like Nancy were more likely to find they could not forget their experience of light. They had seen what happened when whites projected their guilt onto others, and, because they had, they were ready to accept both qualities in themselves. That was what it meant to be able to receive the word. And because Faulkner understood this, he was ready for a journey to a South very different from the South of his youth.

It was not that Faulkner was prepared to abandon the South of *Sartoris* or *The Sound and the Fury*; one still had to swallow the darky and the exotic in Nancy to get at more important features of her characterization. What had happened was that Faulkner had found a way to separate himself from that South without leaving it. "That Evening Sun" asked what would happen if that South were gone, and, because it did, this story was subtle at exactly the points where his novels had not been. Faulkner was ready for one of his more remarkable achievements, one that would liberate him, for the moment, from a great many of the ties that bound him to the world that with *Sartoris* he had feared "to lose and regret."

II

Light in August (1932) began with a poor white girl, Lena Grove. Pregnant and deserted, Lena had set out on foot to find her lover, and as she gravitated from one Puritan community to another she was treated as a pariah because of her condition. But Lena was oblivious to this reaction. She had a special quality of faith, alien to the Puritans around her, which overcame their prejudice and inspired their help. The novel's title, Faulkner said, referred to Lena and what she stood for. In Mississippi in August, he explained, "there's a few days . . . when suddenly there's a foretaste of fall." At such times he had observed "a lambence, a luminous quality to the light"; it reminded him of "a luminosity older than our Christian civilization." Such a light might have something "from Greece, from Olympus in it somewhere," and "maybe," he added, "the connection was with Lena Grove, who had something of that pagan quality of being able to assume everything." That was why Lena "was never ashamed of that child whether it had any father or not"; Lena "didn't especially need any father for it, any more than the women . . . of whom Jupiter begot children were anxious for a home and a father."[10]

In Lena Faulkner introduced a woman who by some miracle had never been indoctrinated by the Puritan whites around her. Her light was human, not divine, so she could "assume everything"; accepting both the good and the evil in herself, Lena had no repressed guilt to project onto anyone else. She "seemed to me to have had a very fine belief in life," Faulkner said. But there was going to be much more to *Light in August* than Lena. Lena, with her pre-Christian "luminosity," believed "in the basic possibility of happiness and goodness"; but Faulkner also wanted to dramatize the antithesis of that. He wanted to create someone who had "accepted a tragic view of life,"[11] someone who was polarized between the extremes of light and dark, as whites like John McClendon and Minnie Cooper were. But now, instead of looking toward whites, he turned toward blacks.

In "That Evening Sun" the black man became the "obverse reflection of the white people he lives among" only when he accepted white stereotypes about his identity. But what if a man like that had no sense of ethnic identity to fall back on? Like most southerners, Faulkner had known blacks who could have passed for white; freckled Chess Carothers, J. W. T. Falkner's driver, probably could have passed, as could Oxford black community leader Rob Boles.[12] But what Faulkner now proposed went far beyond passing. What if such a man were raised among unsuspecting white Puritans in the belief that he had black ancestry, yet were somehow cut off from all meaningful contact with blacks? To the degree that such a man thought of himself as black, he could never be any *more* than the "obverse reflection" of whites. Since he would never know any other black identity than the demonized one he inherited from the Puritans around him, he would be polarized between his public white identity and his secret black one. He would not be able to conceive a life in which guilt might not fall on him any time his identity was discovered. What was worse, his training as a Puritan male would keep him from accepting any guilt, however justly that guilt might be assessed. The razor-carrying black Jesus of "That Evening Sun" had been an ironic symbol of the Word made flesh because he stood for the powers of light; Joe Christmas, the "parchment-colored" (*LIA*, 91) razor-murderer of *Light in August*, would be an ironic symbol of flesh without the Word because he could never identify himself with either. Such a man might suggest little about the kind of ethnic identity blacks inherited from each other, but he would "isolate white facts" about prejudice, revealing dramatically the human tragedy behind Puritan scapegoating.[13]

Writing about poor whites opened new vistas for Faulkner. While writing about aristocrats, he had been unable to separate himself from the bland racism he had inherited with his class; but now, writing about poor whites, he seemed liberated. The way he conceived his new hero showed how that had happened. Faulkner had not forgotten the racist propaganda that had set the tone of the class struggles in the world of his youth; his new hero was going to be the victim of racists like James Kimble Vardaman, who thought that the black was a "lazy, lying, lustful animal" whose nature "resembles the hog's." Their power over him would be so complete that he would think he was the devil-nigger they claimed he was; he would commit a crime as bloody as the one Nelse Patton had committed and become the apotheosis of all their Manichaean fears. Through him Faulkner would evolve a sweeping vision of poor-white racism. Rednecks, Faulkner had discovered, were racists because they were Puritans, and he was going to put the blame for that racism where it belonged. His vision, moreover, was going to transcend regional boundaries, and to emphasize that point he decided to reveal it through a family of New England abolitionists who had settled in Jefferson after the Civil War.

The last descendant of this family was Joanna Burden, a spinster who would become Joe Christmas's lover and who would die at his hand. Joanna told Joe how Nathaniel Burden, her father, had taken her as a child to a cedar grove where the two carpetbaggers whom John Sartoris had killed on election day in 1872 were buried. The carpetbaggers were Joanna's brother and grandfather; Nathaniel was trying to help Joanna understand why they had died and to realize that what had killed them was not merely John Sartoris. Their killing had been preordained, he said, "before your grandfather or your brother . . . were even thought of"; it was the result of a "curse which God put on a whole race." The race he meant was not his own white race, but the black race, and here Nathaniel turned to the apocryphal story of Cain. "The curse of the black race is God's curse," he said; but in Nathaniel's version, Cain's curse had been visited on blacks because God wanted to punish not blacks, but whites. His Reconstruction experiences, Nathaniel said, had shown him that "the curse of the white race is the black man," who was "doomed and cursed to be forever . . . a part of the white race's doom and curse for its sins." That meant that one could not avoid the black man's curse by being white, for God's curse was the "curse of every white child that ever was born and that ever will be born" (*LIA*, 187–88).[14]

This speech of her father's had been traumatic for Joanna. It had changed the way she thought about blacks. "I had seen and known negroes since I could remember," she told Joe, and she had taken them for granted. "I just looked at them as I did at rain, or furniture, or food or sleep." But Nathaniel had opened up a nightmare world to his daughter, and she began to conceive a fantasy about blacks:

> I seemed to see them for the first time not as people, but as a thing, a shadow in which I lived, we lived, all white people, all other people. I thought of all the children coming forever and ever into the world, white, with the black shadow already falling upon them before they drew breath. And I seemed to see the black shadow in the shape of a cross. And it seemed like the white babies were struggling, even before they drew breath, to escape from the shadow that was not only upon them but beneath them too, flung out like their arms were flung out, as if they were nailed to the cross. I saw all the little babies that would ever be in the world, the ones not yet even born—a long line of them with their arms spread, on the black crosses (*LIA*, 187–88).

Nathaniel Burden was conducting his daughter through the rites of transition, passing on the Manichaean Puritanism he had inherited. "You must struggle, rise," he told her. "But in order to rise you must raise the shadow with you." In the Reconstruction South, Nathaniel admitted, he had learned that "you can never lift it to your level." Traditional Christians knew Jesus was sent to resolve that dilemma, but Nathaniel could see only a struggle in which whites were predestined to fail. God had two hands, and Nathaniel thought that the black man would "be forever God's chosen own because He once cursed Him" (*LIA*, 188).[15]

Nathaniel's tale showed the rituals by which Puritans, northern or southern, identified themselves in *Light in August*. They projected the powers of light onto the Puritan male so that he could control the darkness they projected onto everybody else; the Puritan home was the institution through which they celebrated this ritual. The story of Joanna's family illustrated the process. Nathaniel's father Calvin, a "tall, gaunt Nordic man," ran away from his New England home at twelve, shipped around the Horn to California, wandered overland to Missouri, and took a wife, a short, dark Huguenot named Evangeline. For this illiterate son of a Puritan minister, revolt meant wandering to the south, associating with Latin races; home meant returning to the north, marrying. Calvin wanted his children "to hate two things" that reminded him of his youthful revolt and southern wanderings, "hell and slaveholders"; he meant to project these onto short, dark Evangeline and control them by controlling her. Predictably, Calvin's son Nathaniel ran away at puberty, too, also southward, to Mexico—and when he returned in 1866 with a short, dark girl named Juana who had borne him a son, the old man was frightened to see history repeating itself. Dark Nathaniel, "with his beaked nose and white teeth," did not look at all like "Nordic" Calvin; father and son seemed "like people of two different races," and to Calvin his grandson suddenly looked like an omen. The boy had a "black look," he complained, because he had a "black dam" (*LIA*, 179, 180, 182, 184).

Calvin's complaint identified his family's vision of American life. It was Satan, after all, whom Puritans knew as the Black Man. For Calvin the curse of Cain meant that people who owned black men were in league with Satan. The Latin races, he was aware, had been the first Western Europeans to own black slaves, and he knew they were the first to bring them to America. That was why, when the Burden men rebelled, they went south where the Latin peoples were: running away from home meant running from God. Calvin was sure the pigmentation of the Latin races was the result of the devil's bargain of slavery. "Damn, lowbuilt black folks," he ranted. They were "low built because of the weight of the wrath of God," and they were "black because of the sin of human bondage staining their blood and flesh" (*LIA*, 183).

For Faulkner the Burdens were the same kind of Manichaean Puritans John McClendon, Minnie Cooper, and Stovall were. They were racists, and racism was the paradoxical motive behind their commitment to abolition. These Puritans cast themselves as Nordic exemplars of light, absolving themselves of their guilt by projecting it onto women, onto children, and onto all races whose pigment was darker or whose origins were more southerly than their own. Mexicans, French people, Catholics, and southern slaveholders all found their places in a Puritan spectrum in which the Puritan set himself at one end and the African at the other. Any point between represented a stage of miscegenation and signaled a league with the darker powers.

That was why for Calvin Burden the Emancipation Proclamation was a personal victory; it meant an end to this devil's bargain. It meant that black slaves and the white aristocrats who held them would both be free; both classes could exorcise their blackness now, and Calvin thought they would do it the same way Evangeline and Juana had—by marrying Nordic Puritans. When Calvin looked at his own descendants, he was afraid his family was degenerating; but what Calvin had in mind for southerners, "both black and white alike," was the reverse of that process. God's curse, Calvin thought, was going to be lifted through a kind of benign miscegenation. "They'll bleach out now," he said. "In a hundred years they will be white folks again. Then maybe we'll let them come back into America" (*LIA*, 183). History was a whitening process through which the darker races were supposed to "bleach out," and the Puritan home was the instrument through which this whitening was to be achieved.

Faulkner had gone to the wellsprings of Anglo-American folk tradition—New England—to trace the origins of that Manichaean Puritanism; but his vision was a sweeping one that embraced all of Puritan America. Its pervasiveness became obvious the night before Joe Christmas was killed. Defrocked southern minister Gail Hightower, whose aristocratic origins gave him a certain perspective, sat by his window as music rose from the nearby church that had disowned him. "Listening," Hightower seemed "to hear within . . . [the music] the apotheosis of his own history, his own land, his own environed blood." In this music Hightower felt "a quality stern and implacable, deliberate." It was very different from the music of the black singers of *Soldiers' Pay*; it was "without passion so much as immolation." It pleaded "for not love, not life, forbidding it to others, demanding in sonorous tones death as though death were the boon"; and it was "like all Protestant music." Hightower had sprung from a people who were uneasy with goodness. "Pleasure, ecstasy, they cannot seem to bear," he decided, and he saw that "their escape from it is in violence." To Hightower the violence consisted not only in "drinking and fighting" but also in "praying," and these thoughts forced a conclusion: "*And so why,*" he asked himself, "*should not their religion drive them to crucifixion of themselves and one another*?" (*LIA*, 273).

The Burdens' God of light had made the darkness His "chosen own" because he was no Christian God. The children Joanna saw in her fantasy were not going to find redemption. Joanna's father had been right: those children could not raise the shadow to their level, and for that reason they could never ascend toward the light themselves. During life the lighter and darker natures of those children would be locked in an endless struggle neither could win. Only death could set them free, and that meant being crucified on the black cross. The terrible black cross was really a symbol of freedom! People like the Burdens learned to love the darkness the same way their God did, because they knew that only the black cross could free them. That was why the Burden men wandered south, ready to confront the darker races, why the family's fate was already sealed when they moved from Missouri to Jefferson; they were drawn there by their Manichaean faith. It was there, Joanna said, that "what they probably knew all the time was going to happen did happen" (*LIA*, 186): they were killed by John Sartoris, who as a former slaveholder stood for all the evils in their demonology.

Writing about Puritans had served Faulkner well in his search for a South; it had provided scope for his nostalgia for the world of the Sartorises and Compsons and allowed him to create a subtle myth that explained what had happened to them. Most people of Faulkner's time thought that the conflicts between the North and South, and those between southern aristocrats and rednecks, were separate; Faulkner wanted his readers to know that what had happened to the South was more subtle. The conflict was really between Puritans and Cavaliers, and had been since the seventeenth century. The Cavalier ethic might be decadent, but what the Cavaliers had been fighting was vicious. Puritans like the Burdens and their southern counterparts were zealots and, worse yet, they were not really Christians. If the Cavaliers were decadent, Faulkner now seemed to say, it was no wonder, for their class had been battered by this conflict for three hundred years.

Faulkner had not shaken himself free from the South of his youth. This new myth he had created seemed to exonerate the Sartorises and Compsons. It suggested that the planter and slave were victims of the same Puritan prejudice and that in the face of such prejudice the aristocrat was right when he said he was the main defender of blacks in America. The myth shifted the blame for both the Civil War and the revolt of the rednecks from the shoulders of the aristocrats and placed the blame for slavery on the very people who had complained loudest about the Cavalier's tyranny.

The irony was that by shifting the blame Faulkner found himself strangely liberated. Relieved of the burden of defending his own class, he could write about the South's two plebeian classes with a freedom he had never found before; and he had learned how to write about them without penetrating the masks that had frustrated him as he struggled with Dilsey and the black singers of *Soldiers' Pay*. Joe Christmas, raised as a Puritan white with the secret belief that he had black ancestry, was no black American in the ordinary sense of the term; his character, formed in isolation from all significant contact with blacks, suggested little about a black ethnic heritage. But it did suggest a great deal about prejudice and its causes. Depriving Joe of an ethnic heritage served to "isolate white facts" in the same way Ellison did three decades later when he asserted that blacks were "caught up associatively in the negative side of . . . [a] basic dualism of the white folk mind." And it showed Faulkner the materials of a genuine black American tragedy. The irony of *Light in August* was that Joe Christmas, whose self-image was formed almost entirely by white Puritans, was going to be the most suggestive "black" character in Faulkner's canon, launching Faulkner on a journey to a South very different from any he had visited before.

6

How to Visit the Black South without Visiting Blacks

I

As a young man riding a Jim Crow car to Chicago, Richard Wright pondered his Mississippi youth. During his formative years, Wright concluded, "there had been slowly instilled into my . . . consciousness, black though I was, the culture of the South." For Wright understanding himself meant understanding the South; he was not leaving the South, he was only "taking a part of the South to transplant in alien soil, to see if it could grow differently."[1] Wright transplanted one part of that South into *Native Son* (1940), and his hero, Bigger Thomas, showed blacks a new way of thinking about themselves. This ghetto youth, given a chauffering job by Chicago millionaire Henry Dalton, was helping his employer's drunk daughter Mary to her room when blind Mrs. Dalton entered; he panicked and put a pillow over the girl's face to keep her from attracting attention; she smothered. As Bigger sat in a streetcar after disposing of Mary's body, this once cowardly youth felt a strange exhilaration: "Would any of the white faces all about him think that he had killed a rich white girl?" Bigger was fascinated. "No! They might think he would steal a dime, rape a woman, get drunk, or cut somebody; but to kill a millionaire's daughter?" At that thought Bigger "smiled a little, feeling a tingling sensation enveloping all his body."[2]

In 1955 James Baldwin attempted to explain Bigger's reaction. All too often blacks came through to white Americans merely as symbols of their white guilt, Baldwin thought; it was in "reaction" to this guilt that the "image of Bigger was created." That image, he said bluntly, was the " 'nigger,' black, benighted, brutal, consumed with hatred as . . . [whites] are consumed with guilt." Whites, not Wright, had created that image; but Baldwin was sure that the "American image of the Negro lives also in the Negro's heart": "there is, I should think, no Negro living in America who has not felt . . . naked and unanswerable hatred; who has not wanted to smash any white face he may encounter . . . , to violate, out of motives of the cruelest vengeance, their women, to break the bodies of all white people." Every black made a "precarious adjustment" to both sides of that image, Baldwin felt: "to the 'nigger' who surrounds him" — the fear whites felt because of the guilt they projected onto him — and "to the 'nigger' in himself" — his own feelings of hatred and vengeance. Sometimes, when that adjustment got out of kilter, a man was gone: "when he has surrendered to this image life has no other possible reality." That was what had happened to Bigger; he had become "a monster created by the American republic."[3] And it explained why Wright, as he struggled to articulate the meaning of his "monster," was ready to assert that Americans had "in the oppression of the Negro a shadow athwart our national life dense and heavy enough to satisfy even the gloomy broodings of a Hawthorne."[4]

Nine years before Wright showed what his transplanted South had done to Bigger,[5] Faulkner had been ready to visit a South that sounded remarkably like the one Wright had escaped. For Faulkner it was a strange South, where blacks and Puritan whites struggled without the restraining hands of the aristocrat to protect them from each other. Learning about that South would show Faulkner, as nothing else would, the value of the aristocratic tradition that seemed to be dying all around him; it would also show him something he did not expect to find. Colonel W. C. Falkner's great-grandson was about to awaken to a perceptive concern that travestied his mellow nostalgia for the South of his youth; and he was going to do that precisely because *Light in August* allowed him, for the moment, to burrow under all the superstrata of paternalism that he had inherited from that South.

Aristocrat Quentin Compson had learned that what whites called "nigger" behavior was an "obverse reflection of the white people . . . [a black] lives among" (*SF*, 67). There were things blacks learned about a black identity from each other, but what whites saw tended to be things the blacks learned about from whites. What if a man were raised among unsuspecting white Puritans, secretly believing himself to be a "nigger" — believing, that is, that he had inherited the proverbial drop of "blood" that was supposed to pollute the whole? That would isolate what blacks learned about their identity from whites. Such a man would have no knowledge of what it meant to be black except what the whites taught him; in Baldwin's terms, "the 'nigger' who surrounds him" — the image of whites' fears — would *be* "the 'nigger' in himself." The potential here for identifying how white prejudice shaped black behavior was almost infinite. Joanna Burden's fantasy of the black cross was a vision of what Wright called the "shadow athwart our national life"; "parchment-colored" Joe Christmas was the apotheosis of that shadow, the "monster created by the American republic."

Faulkner conceived the story of Joe's birth as a fable of the birth of such a "monster." Doc Hines, Joe's grandfather, was a southern Puritan whose apocalyptic racism resembled that of the Burdens. Doc was less patient than Nathaniel Burden, who thought southern whites might be freed of the shadow of the black cross in "a hundred years" through a kind of benign miscegenation that would allow them to "bleach out": Doc preached black genocide. As wild as the Burdens when he was a youth, Doc had been in jail when his daughter was born; overcome with guilt, he had evoked the Puritan ritual of casting himself as defender of light, projecting that guilt onto his daughter. When this girl grew up, it sounded like the Burdens all over again: she ran off with a dark-complected man she insisted was Mexican. Doc responded in Manichaean terms. He could "see in his face the black curse of God almighty," he proclaimed—and he killed the man. "I have brought you back the devil's laidby crop," he told his wife when he took the pregnant girl home. No wonder his saintly wife had a premonition of the Manichaean tragedy that lay ahead: "Sometimes," she said, "I would think how the devil had conquered God" (*LIA*, 183, 278, 280).

Doc's God was the same God the Burdens thought had made the darkness His "chosen own," and this God had told Doc to "watch My will a-working." That, Doc was sure, meant to watch "for evil to come from evil." But Doc was not content to wait for God; when his daughter gave birth to a "parchment-colored" boy, he began a complex charade designed to teach his grandson what blackness meant. He left Joe on the doorstep of a white orphanage; then, concealing his identity, Doc took work at the orphanage boiler room. When Joe was old enough to understand, Doc hinted to the boy that he carried the mark of Cain. "Do you think you are a nigger," he asked, "because God has marked your face?" Other children picked that up and began to call Joe "nigger" (*LIA*, 188, 284, 286, 285). Joe's early sense of self, formed at a time in early childhood which he could not later recall, reflected Doc's Manichaean notion of evil. In Baldwin's terms, "the 'nigger' in himself" reflected "the 'nigger' around him": the shadow of his white grandfather's guilt. Baldwin's malediction that the "American image in the Negro lives also in the Negro's heart" was in Joe's case literally true; he was already on his way toward becoming the "monster created by the American republic."

To this point Joe knew nothing about what being black meant to blacks, but he was soon destined to learn the kind of problems white society forced on them. "In our society," Ralph Ellison wrote in 1944, "it is not unusual for a Negro to experience a sensation that he does not exist in the real world at all." A man like that, Ellison thought, might seem to himself only "to exist in the nightmarish fantasy of the white American mind" as the image of its guilt: "a phantom that the white mind seeks unceasingly . . . to lay."[6] If he was black, Joe decided, he had better learn about being black from someone who was black; he asked a black workman, "How come you are a nigger?" To Joe, the question made sense; if a man was a "nigger," it must be for a reason. But this workman, angry at such a question from a "parchment-colored" child, gave Joe more truth than he was ready for. "You are worse than that," he told the boy. "You dont know what you are. . . . You'll live and you'll die and you wont never know." "Memory," Faulkner said of Joe, "believes before knowing remembers. Believes longer than recollects, longer than knowing even wonders" (*LIA*, 285, 88). If belief, in some matters, came prior to understanding and was impervious to it, Doc's Manichaean teaching had already made Joe a "nigger"; there, back behind everything Joe could recollect, where he could never get at it cognitively, was the identity problem with which white society infected blacks. Joe's tragedy, Faulkner said later, was that "he didn't know what he was, and so he was nothing." That, Faulkner believed, was "the most tragic condition a man could find himself in."[7]

Doc had bequeathed Joe another legacy: he had shackled him to a social identity marked out in American society as that of the scapegoat. White Puritan society projected its guilt onto blacks, women, and children and punished them in the name of God; but being a "nigger" was the only identity that the orphaned Joe had, and so Joe's identity was his guilt. Being guilty for what one was born was terrible, a kind of psychological castration, and Joe was soon to learn that he was to spend the rest of his life running from such punishment. Simon McEachern, the white Puritan farmer who adopted Joe, seldom talked about race, and he had no notion that Joe's father might have been black; but with no blacks to project his guilt onto, he projected it, as other Puritans did, onto members of his family. Joe, who secretly believed he was black, took every effort McEachern made to manage him as an attack on the "nigger," which was the only identity he had. The confrontation came when McEachern threatened to whip him for not memorizing his catechism. For eight-year-old Joe, the stakes were clear: the price of yielding would be his manhood. It was one thing to suffer the moral castration of being shackled to a "nigger" identity and punished for it; but being lashed naked with a leather strap was all too much like a real one. Joe was not yet ready to fight back, but he was ready to assert his identity. "He stood

. . . erect" as the blows fell, Faulkner reported, and later he would remember that *"on this day I became a man"* (*LIA*, 111, 108).

Joe was right; after that moment, there was one part of him nobody could touch. But the problem was more complex. Faulkner wanted to make a place in his fable for what Baldwin spoke of as the "perverse and powerful desire" blacks felt "to force into the arena of the actual those fantastic crimes of which they have been accused." Beatings changed a person however he responded to them. To affirm the manhood Joe feared these beatings were destroying, the boy began to do what McEachern had forbidden him to do. Joe began to enjoy sneaking. He stole from Mrs. McEachern's meager savings. Climbing out the window at night, he gravitated to a café that he did not know was a bordello and fell in love with Bobby Allen, a waitress he did not know was a prostitute. McEachern thought he was training Joe to avoid those things by beating him; but he was really training him to enjoy them. It was hardly surprising that when Joe found out about Bobby he accepted her anyway, connecting her guilt at being a whore with his secret guilt at being black.

Such was Faulkner's parable of the way pressures from white society shaped black character. If a black, as Baldwin said, had to make a "precarious adjustment" to his feelings of hatred and his desire for vengeance ("the 'nigger' in himself"), it was because those feelings were a mirror of the fear and guilt whites projected onto him ("the 'nigger' who surrounds him"). But there the process reversed itself and pointed back toward whites, reinforcing that fear. Flouting his foster-father's rules made Joe feel real; but Joe's conduct only affirmed McEachern in his belief that Joe was controlled by Satan. For men like McEachern, Faulkner reported, "bigotry and clairvoyance were practically one," and, when he found Joe with Bobby, he did not see "the face of the youth whom he had nurtured and sheltered . . . from a child." That was because "it was not the child's face which he was concerned with: it was the face of Satan, which he knew as well" (*LIA*, 148, 150, 151).

Although McEachern had no idea that Joe might be black, his responses were little different in quality from those of James Kimble Vardaman or of the Oxford men who had lynched Nelse Patton in 1908, and these responses confronted the youthful Joe with the same problem blacks faced. A mature black might separate the need to declare his manhood from his desire to strike back at the society that was sapping it; but not everybody who faced Joe's problems was that mature. "You ache to convince yourself that you do exist in the real world," Ellison's invisible hero explained, "and you strike out with your fists . . . to make them recognize you"; insulted on a dark street, he was shocked to find himself in a murderous "frenzy"; he beat the blonde stranger to the pavement, and "in my outrage I got out my knife and prepared to slit his throat."[8] That was what Baldwin meant when he said that the "American image of the Negro lives also in the Negro's heart." There were times when, to keep his manhood, a black man had to be the monster whites thought he was. He was forced to violence because then, "like the white enemy" he was struggling with, he had "no means save this of asserting his identity."[9] Joe now became the devil-nigger he saw in McEachern's eyes; erupting in a rage, he felled his foster-father with a chair, killing him. For Joe, as for Bigger Thomas, there was no remorse, only the thrill of existential commitment. "I said I would kill him some day!" he cried and rode away, "exulting perhaps . . . as Faustus had, of having put behind now at once and for all the Shall Not." And that commitment was complete. "Get away, old woman," Joe told Mrs. McEachern as he ran upstairs to steal the rest of her money, and he wanted her to know how important it was that he was stealing it. "I didn't ask you for it," he said. "Remember that. . . . I just took it" (*LIA*, 152, 153, 154).

To gain his manhood, Joe had had to assume responsibility for evil as well as for good, but he was not able to separate that responsibility from his repressed desire to strike back. That meant commitment to a blackness that was inherited socially, not genetically. As Joe ran upstairs after the money, Faulkner reported, he seemed to "be vanishing as he ran." To Mrs. McEachern it was "as if he were running headfirst and laughing into something that was obliterating him" (*LIA*, 153). To his foster-mother Joe had always been invisible; it would have made little difference if he had vanished into light instead of darkness. He was lost, and Faulkner's later explanation of how that had happened sounded remarkably like the kind of explanations Ellison and Wright gave for their characters. Joe knew that "in order to live with himself," Faulkner said, "his only salvation . . . was to repudiate mankind, to live outside the human race." Joe "didn't know what he was, and so he was nothing."[10]

II

Bigger Thomas's change from Tom to devil-nigger might have shown him how to be a man in a white male world; but if Bigger had learned something about that, he appeared to have done it in spite of the women in his life. It was his pious mother and nagging sister who had driven Bigger into the street; Bessie, his black girl friend, gave him no more than drunken sex; and white Mary Dalton, who wanted to befriend Bigger, made him feel like a Tom. Even the thrill of commitment Bigger gained had not come from confronting a white man; it had come from accidentally violating white taboos when he killed a white woman, and his motivation had been fear of still another woman, her mother. Wright never developed those implications of his tale; but, to Faulkner, the effect black and white women had on black manhood seemed as important for Joe's manhood as his relationship with white males.

The murder of his foster-father would have repercussions in Joe's relationships with women. McEachern was largely responsible for forming Joe's image of white men, and in killing him Joe was rejecting conscious identification with this image of manhood; he would identify himself in opposition to this image now, through the kind of violence whites associated with blacks. But if Joe were going to have any kind of nonviolent identity, he had to find it through his relationships with women; he had to find a place for a black man in the Manichaean ritual that had white males projecting their guilt onto blacks, women, and children and controlling them in the name of God.

Again, the roots of Joe's problem were in his early childhood. Scrounging in the dietitian's room at the orphanage one day, the boy began to eat from a tube of toothpaste. When the dietitian and her boy friend came in and began to make love, Joe had no idea what they were doing; but, when he vomited toothpaste and was discovered, they assumed he did. "Spying on me!" the woman spat. "You little nigger bastard!" (*LIA*, 91). But then, realizing her reputation was in the boy's hands, she tried to bribe him not to tell. That was Joe's first experience of what sex meant to white women. Joe had no idea what these two adults were doing, but he had been caught sneaking the toothpaste, and he expected to be punished; instead, he was rewarded. That, Faulkner seemed to be saying, was what happened between white women and black boys, and, for Joe, the damage was going to be permanent; at the back of his consciousness, imprinted at a time he could scarcely remember, the set of ideas he connected with white women would include fear, guilt, nausea, money, being called "nigger."

Living with the McEacherns had reinforced all these ideas. Pious Mrs. McEachern resembled the Burden women. She came through as Faulkner's typical Puritan woman: timid, "a little hunched, with a beaten face," the antithesis of her domineering husband. Puritan males separated themselves from their guilt by projecting it onto others and controlling those others in the name of God; Puritan women accepted their guilt and projected their desire for salvation onto their domineering men. It was "as if she were the medium," Faulkner reported, "and the vigorous and ruthless husband the control." Equal in quantity to her husband's,

Mrs. McEachern's guilt was different in quality, and for Joe that would be devastating. Bad as McEachern was, he "merely depended on . . . [Joe] to act in a certain way and to receive the as certain reward or punishment." But his foster-mother had "a woman's affinity and instinct for secrecy," Joe felt, "for casting a faint taint of evil about the most trivial and innocent actions." McEachern was trying to beat Joe into a castrated, Tomish self-image; Mrs. McEachern was trying to seduce him into one. Joe understood that the day he refused the catechism. As the boy lay beaten in his room, his foster-mother brought food to him, but there seemed to be no love in the act; it was as though woman and child were automatically involved in some dark conspiracy against the man. "He never told me to bring it to you," she insisted (*LIA*, 109, 123, 124, 114). To Joe, accepting that food seemed as frightening as being beaten; it meant acknowledging fellowship with Mrs. McEachern as objects of McEachern's control over the darkness.

This female attack was also harder to respond to. McEachern had engaged Joe in a battle of strong wills, but Mrs. McEachern appealed to his weakness. As Joe struggled to find a response, he set a pattern that remained with him: seizing the tray, he dumped plates and food onto the floor. Joe had meant to respond to his foster-mother the same way he had responded to McEachern, by acting the devil-nigger instead of the Tom she was trying to make him; but he was already beaten as no whipping could have beaten him, and when she left he surrendered: he ate from the floor "like a savage, like a dog." Male violence he could deal with; what Joe learned to fear most in his youth was "the woman: that soft kindness which he believed himself doomed to be forever the victim of" (*LIA*, 115, 124).

There was no middle ground. Joe needed to break out of that Manichaean framework altogether, and that was why, at seventeen, he gravitated toward the prostitute, Bobby Allen. Bobby offered hope. Here, it seemed, was a white woman outside the Puritan framework who could love Joe for himself. "I got some nigger blood in me," he confided to Bobby (*LIA*, 144), and, when she refused to acknowledge that he might be black, Joe thought he had found something like acceptance. But those Puritan sex rituals, Joe found, pervaded everything; what went on in a whorehouse was only a variation of the same Manichaean theme. Projecting his guilt onto whores instead of children, blacks, or other women, the Puritan male separated himself from his guilt by giving them money instead of dominating them. That was why Joe sensed Bobby's alienation was like his own: the role forced on prostitutes as the price of survival was similar to that forced on black males. The whore struggled to keep her womanhood in the face of such rituals; the black male struggled to keep his manhood. Both were more concerned with surviving than with transcending those rituals, and Joe's discovery that Bobby was a whore meant that he was already fatally caught.

Bobby had taught Joe what passed for a white male role. When she accepted the petty sums Joe stole from Mrs. McEachern, Bobby thought she was allowing Joe to project guilt onto her and separate himself from it by paying her. That, not acceptance, was Bobby's reason for ignoring Joe when he told her he might be black; as long as Joe was supposed to be white, the ritual could continue. But Bobby had another motive, too, and it surfaced after Joe killed McEachern. Now it was Bobby who flew into a rage. "Bastard!" she screamed. "Getting me in jam, that always treated you like you were a white man!" (*LIA*, 160). Since his confession, Joe had never been anything but a "nigger" to Bobby. Taking what Joe could pay her, Bobby had really been losing money she would have made with other clients; in effect, she was paying Joe, and Bobby knew what she was doing. Convinced Joe was black, she could project her guilt onto him and exorcise it through his failure to pay. Bobby was a Puritan whore, and, when she and her friends left Joe beaten into unconsciousness, he was at a turning point in his life. Hatred, violence, castration, the passage of money: those were the ways whites in the sanctified world of Simon McEachern reacted to a black identity. Now Joe knew that that was true in the world of Bobby Allen as well.

Joe had little reason to think those conditions would change for him; but Bobby had taught him better than she knew. Youthful Eldridge Cleaver, out of jail after a marijuana conviction, had explored ways of searching out a sexual identity through violence against white women. For years Cleaver had been having masochistic fantasies about white women; but if whites were the enemy, he now decided, why not turn those fantasies around and act them out? "I became a rapist," Cleaver confessed. "I did this . . . deliberately, willfully, methodically." Cleaver was projecting his own guilt over those fantasies onto white women and exorcising it by dominating them, and he felt something like the strange thrill of revolutionary commitment Bigger Thomas felt after killing Mary. "It delighted me that I was defying and trampling upon the white man's law," he remembered. "I felt I was getting revenge."[11]

Joe was not yet ready for rape, but he had learned there were more ways than love to create a male role with white women. When Joe had money, he could pass for white and project his guilt onto whores the way white men did, and, when he could not pay, the experience with Bobby had showed him another role: he "bedded anyway and then told them that he was a negro." That seemed to work just as well as money: "It was quite simple, quite easy," although sometimes Joe had to take a beating from the "other patrons" (*LIA*, 166). Joe had learned how to turn one Puritan ritual to his advantage. Those whores, he had learned, wanted an outlet for their guilt as much as they wanted to get paid; their projections affirmed Joe's male role as devil-nigger, and, when he was beaten by the "other patrons," their rage affirmed that role, too.

Learning to turn those rituals back on the whites who practiced them was a lesson that would stay with Joe, but it was a sick business. Cleaver, remembering his rapes, was aware of a profound desperation just beneath the surface of his consciousness: "looking back I see that I was in a frantic . . . and completely abandoned frame of mind."[12] Establishing a self-image through hatred dehumanized a man as effectively as castration; and what was worse, he was still dependent on whites for that self-image. Joe's desperation surfaced when he found a whore who was indifferent to Puritan rituals. "What about it?" this northern woman asked. "You look all right. You ought to seen the shine I turned out just before your turn came." Joe "did not know until then," Faulkner reported, "that there were white women who would take a man with a black skin" (*LIA*, 166, 167). For Joe, that was devastating. This woman was refusing him his male devil-nigger role, and the only role left for him was the castrated, Tomish one; he was face to face again with that hell of castration Mrs. McEachern's "soft kindness" had shown him. Struggling to salvage the devil role, he beat the woman; but this time, Joe knew, violence was not enough.

Joe needed to take stock. If he was sure of his black identity, why did he have to find it with white women? In Soledad prison, Cleaver had to face that in himself, and the experience was traumatic. Cleaver was mortified when a white guard made him take down a picture of a white pin-up girl: "I realized that I had chosen the picture . . . over the available pictures of black girls." That brought on "a terrible feeling of guilt," and Cleaver asked himself, "Was it true, did I really prefer white girls over black?" Most of his black friends did. Cleaver had to conclude that "as a matter of course, a black growing up in America is indoctrinated with the white race's standard of beauty," but that did not help. His desperation surfaced when he saw a picture of the Mississippi white woman whom Emmett Till, recently lynched, had been accused of making advances to. "I looked at the picture again and again, and in spite of . . . the hate I felt for the woman and all that she represented, she appealed to me." Cleaver "flew into a rage" at the situation: "at myself, at America, at white women." Shortly afterward he had a "nervous breakdown."[13] That was where Joe Christmas now found himself. If he avoided black women and insisted on white ones, if white women were somehow necessary to him for his self-image as devil-nigger, then in a sense he had been "indoctrinated with the white race's standard of beauty." Joe's reaction was similar to Cleaver's. "He was sick after that," Faulkner reported (*LIA*, 167). When Joe recovered, he was ready to face what he had been avoiding.

Joe's notion of what being black meant had been what he had learned from whites; now he was ready to find out what blacks thought it meant. Joe had been avoiding that knowledge since puberty. At fourteen, he and several white boys had lured a black girl into a shed. When Joe's turn with the girl came, he had been shocked to find himself sick—as he had felt "when he . . . used to think of toothpaste" (*LIA*, 115). But there had been more coming up from Joe's unconscious than his memories of the dietitian who called him a "little nigger bastard," and Cleaver's experience suggested what that might be. "I love white women and hate black women," one black admitted to Cleaver in Soledad. "It's just in me, so deep that I don't even try to get it out of me any more." This man felt "something delicate and soft" in a white woman, he said. "But a nigger bitch seems to be full of steel, granite-hard and resisting."[14] That was something more than being "indoctrinated with the white race's standard of beauty"; it was being indoctrinated in a notion of ugliness, too, and it showed what fourteen-year-old Joe had gone through. Joe had felt repelled by the girl, Faulkner wrote, because he thought he was "smelling the woman, smelling the negro all at once." If that had been a white girl in the shed, Joe could have affirmed his maleness by projecting guilt onto her and controlling her; but making love to a black girl was like making love to his own guilt. It meant affirming what he wanted to deny in himself, and the roles whites projected onto women and blacks suddenly ran together in Joe's mind. He felt "enclosed by the womanshenegro"; bending over the girl, "he seemed to look down into a black well." Joe's experience with that girl was a vision of the very pit of hell, and in its black depths he seemed to see clearly what hell meant: castration. At its bottom her eyes were "two glints like [the] reflection of dead stars" (*LIA*, 115). Joe beat the girl, but the vision would not go away.

That was what Joe was facing in himself, and it was no wonder he had been avoiding blacks; but now he was ready to join them. "He lived with negroes, shunning white people," Faulkner wrote. Some of his identity rituals were familiar. Joe was still violent. Before, struggling to establish his male devil-nigger image with whites, "he had . . . tricked or teased white men into calling him a negro in order to fight them"; but "now he fought the negro who called him white." Still, black men were not what frightened Joe, and he sought out the blackest woman he could find. Joe wanted "to breathe into himself" what he had learned to think of as "the dark odor, the dark and inscrutable thinking and being of negroes." Soon he was "as man and wife with a woman who resembled an ebony carving"; but Joe had been "indoctrinated with the white race's standard of beauty," and that meant he had been indoctrinated with another white standard, too. As Joe lay beside the ebony woman, "his nostrils . . . would whiten and taughten" when he thought about "the odor which he was trying to make his own." He felt "his whole being writhe and strain with physical outrage and spiritual denial" (*LIA*, 167). Joe still felt as "enclosed by the womanshenegro" as he had at fourteen; for Joe, loving a black woman would always mean living in dread of the black pit he had seen in the girl's eyes. And that was why it remained for a white woman to show him that the black pit waiting for him was in his own consciousness.

Joanna Burden, the spinster who carried on her New England ancestors' tradition by doing social work with Mississippi blacks, had never undergone the period of youthful rebellion her father and grandfather had experienced; but she had inherited her share of that rebelliousness, and for Joanna it came in middle life and took the shape of a desire for a black man. For Joanna with her Manichaean upbringing, this was terrifying: instead of remaining the exemplar of light who saved blacks from the power of darkness, Joanna now risked becoming a part of that darkness. As Joe, circling southward again, stumbled on Joanna's isolated house, the stakes for him were just as great; an encounter with this white woman meant an encounter with the dark pit he had feared in the eyes of that black girl long ago. That was going to be the final irony of Joe's struggles for a sexual role: when he had become Joanna's lover, he was committed at last to the darkness he had feared so long in himself. As he crept into her unlighted kitchen, he looked like "a shadow returning . . . to the allmother of obscurity and darkness" (*LIA*, 170).

Both Joe and Joanna were looking for a sex ritual that would allow a viable identity within the Puritan framework. Their tragedy was going to be that neither could find that with the other. Their affair had three phases, and the first set the tone for the other two. Joe was initially satisfied; then he began to feel a strange confusion over male and female roles. Joanna, raised to assume her father's male role as crusader, "had man-trained muscles and . . . [a] man-trained habit of thinking"; she was acting out the role of the Puritan male who projected his guilt onto those around him. Joanna had given Joe a cabin near the house. Every night she left a meal for him in her kitchen; he would enter, eat alone, and then go up to her bed. But when they made love, Joe discovered that it was "as if he struggled physically with another man." By day Joanna cast herself as God's male representative and worked to uplift blacks; by night she substituted Joe, as black, in the role commonly assigned to females in

Puritan society. Joe was not yet ready to come to grips with this reversal, but it was easy to see what not eating together implied: again, he had been cast in the Tomish role, and, to Joe, that was castration. Coming in for his dinner one night, he had had all he could take. *"Set out for the nigger,"* he raged; struggling to re-assert his devil-nigger role, he threw the food on the floor and left (*LIA*, 174, 176).

After a time Joanna came to his cabin, and the second phase began. Something had happened to Joanna. "She is trying to be a woman," Joe saw, "and she dont know how." To Joanna with her Puritan male orientation, being a woman meant being one with the powers of darkness, and Joe discovered that this "New England glacier" was ready to warm herself at the "fire of the New England biblical hell." If that was what being a woman meant to Joanna, Joe was sure he could accommodate her, and this new phase promised him a male role with which he could live. But it was a fatal role for Joanna. Puritan women purged their darker natures by projecting their desire for salvation onto white males and submitting to them; but Joanna could not do that with a black man. Making love to Joe was like making love to the darkness in herself, and, what was worse, she seemed ready to surrender to that darkness; she made love as though she were "damning herself forever to the hell of her forefathers, by living not alone in sin but in filth." Soon she seemed "completely corrupted," and the sign of it was that she was becoming a Negrophobe: in "the wild throes of nymphomania," breathing heavily as he made love to her, she cried compulsively, "Negro! Negro! Negro!" (*LIA*, 191, 178, 193).

Joe was fascinated: this woman loved the devil-nigger in him, and in the safety of companionship he now had a chance to explore that darkness with her. But he soon discovered this new phase was as sick as the first. The white whores whom Joe had told he was black had hated him while they projected their guilt onto him; that had separated Joe from his own guilt. But this woman was loving him while she projected her guilt onto him, and so neither of them could become separated from darkness. Soon Joe began to feel "like a man being sucked down into a bottomless morass." The signal of what was happening was a new change he felt in their sexual roles. In the first phase, while Joanna was projecting a female role onto Joe, it had been "as if he struggled physically with another man"; now, both of them began to feel their roles were female. After orgasm they found themselves lying in each other's arms "like sisters." Joe "began to be afraid," yet something held him; he had lost the power to resist, and that moral castration he had feared so long now loomed as a reality (*LIA*, 193).

It was nature, at last, that forestalled it, bringing on the final phase. Joanna missed her period, and, unable to admit she was entering menopause, convinced herself she was pregnant. For Joanna, that was devastating. Nathaniel and Calvin Burden's youthful revolts had ended when they married dark-complected women, and Calvin's bride had been pregnant at the time; but for Joanna, life was not going to be quite so simple. "She wants to pray," Joe saw, "but she dont know how to do that either" (*LIA*, 194). What was stopping Joanna was the same problem that had plagued them both through the first two phases. Her revolt over, Joanna had to exorcise her guilt through the same Puritan rituals her father and grandfather had used, but Joanna was the wrong sex. The Burden men had married dark Latin women and controlled them in the name of God; a woman attached to a dark man could not do that, and she had to make her peace with that ritual some other way. To Joe, the scheme she came up with sounded outrageous: she wanted Joe to pray with her, become a Christian, and then study law at a black college. Thus sanctified and uplifted, he could take over her white male role of uplifting blacks; and Joanna, projecting her desire for salvation onto Joe, could be exorcised of her guilt the way a Puritan woman was supposed to be, through her man.

That incredible plan showed how desperate Joanna was. "You must struggle, rise," Joanna's father had told her. "But in order to rise, you must raise the shadow with you." For Joanna, Joe had become that shadow; their affair had allowed her to explore her Manichaean father's malediction as her ancestors had, and now she knew he had been right when he had told her that "you can never lift it to your level." Joanna was trapped. Because they were lovers, Joanna now knew that she and Joe were locked in a fatal contest. If she broke off from Joe without raising him to her level, *she* would remain fallen; that was why she was pleading with Joe to change. But what she was asking of Joe was the one thing he could not give; sanctifying Joe would rob him of the only male role with which he could live. To Joanna that meant her only hope of freedom was the black cross she had dreamed of as a child. "Maybe it would be better if we were both dead," she told him (*LIA*, 188, 206).

Joe was beginning to feel trapped, too. "*Something is going to happen to me*," he sensed. Everything in Joe's past told him he had to dominate white women or be castrated by them. "If I give in now," he reflected, "I will deny all the thirty years that I have lived to make me what I chose to be." If he was not going to deny those thirty years, Joe had only one course left; he was at the point with Joanna where he had been with McEachern years ago. As he sat near Joanna's house with his razor, he found himself speaking of the thing as though it had already happened. "*I had to do it*," he mused. "*She said so herself.*" A few hours later, as the passerby who found Joanna described her condition, it sounded like

Mattie McMillan's at the hands of Nelse Patton in 1908: "Her head," he reported, "had been cut pretty near off" (*LIA*, 87, 197, 207, 66). Joanna had found her black cross, and she was free; but she had achieved her revenge, too. For the man who had loved her and destroyed her, momentarily at least, seemed condemned to live.

III

To Wright, allowing readers to witness the rituals Bigger Thomas was caught up in was not enough; Wright wanted to explain the meaning of those rituals, too, and for that, as Bigger faced a courtroom that seemed like the mob that lynched Patton, Wright created Marxist lawyer Boris Max. Max got Bigger to plead guilty; that way, as Max addressed the judge, he could make the trial a forum for the social processes that had brought Bigger there. Max wanted the people in that courtroom to look at those social processes objectively, and if they were going to do that, he said, they had to get past whatever "rage" or "sympathy" they might feel toward Bigger. The common denominator of both those feelings was guilt. "There is guilt in the rage that demands that this man's life be snuffed out," Max told the court, but, "if one insists upon looking at . . . [Bigger] in the light of sympathy," the same motive would emerge—one would "be swamped by a feeling of guilt so strong as to be indistinguishable from hate." To see what had happened one had to dig deeper. The clue, Max thought, was that although Bigger had killed Mary accidentally, "he accepted the crime"—just as though he had done it on purpose. That should have shown the court that for Bigger, killing Mary was "the most meaningful . . . and stirring thing that had ever happened to him," Max asserted, "because it made him free, gave him the possibility of choice." From that standpoint, Max could even see Bigger's murder as "an act of *creation*." That was why Bigger had been "willing after his crime to say: 'Yes, I did it. I had to.' " For Max, that was the clue to what Bigger was caught up in. Understanding Bigger meant "an unveiling of the unconscious ritual of death in which we, like sleep-walkers, have participated so dreamlike and thoughtlessly."[15]

Ex-Mississippian Wright's lawyer could have been pleading for Joe Christmas. Like Bigger, Joe had found in his crime a way to preserve his manhood, to preserve "all the thirty years that I have lived to make me what I chose to be"; it was why Joe, like Bigger, had been able to say, "*I had to do it.*" Faulkner, like Max, was trying to show that this "criminal" was no more guilty or innocent than society itself, that both men were equally caught up in processes that produced first the "criminal," then his crime, then his inevitable martyrdom. But when all of that was said, there was one problem Joe still had to face; and unlike Max, who wanted society to share Bigger's guilt, Faulkner knew Joe had to face that problem alone. Joe might have preserved his manhood; but he had done that at the expense of his humanity. Joe had "evicted himself from the human race," Faulkner explained,[16] and now, somehow, he had to rejoin it.

Faulkner allowed Joe an experience Wright's ghetto youth could never have encountered. Fleeing a Jefferson posse through open country, Joe woke one morning unexpectedly infected with the peacefulness of dawn. There was a "loneliness and quiet" there, he suddenly realized, "that has never known fury or despair." In such a mood Joe found he could think objectively about his motives. Tranquillity like that, he reflected, "didn't seem to be a whole lot to ask in thirty years," and such tranquillity, he suddenly understood, "was all I wanted" from life. Realizing that about himself changed Joe's attitude toward others, and when he asked a black farm family for food he was shocked to find they were terrified. "They were afraid," he mused. "Of their brother afraid" (*LIA*, 246, 249).

The most important thing about Bigger Thomas, Baldwin thought, was his hatred. A man like Bigger "*wants* to die," Baldwin insisted, because of "the premise on which the life of such a man is based." That premise was one that white society had lodged in him, the premise "that black is the color of damnation"; but whatever its origins, a man like Bigger had to be true to it, and that could mean only one thing—he had to die at the hands of his tormentors. It was "the only death which will allow him a kind of dignity," and it might even offer more: "however horribly, a kind of beauty." That, Baldwin insisted, was what people had to understand about Bigger, because it showed the real significance of the devil-nigger role. A man like Bigger "glories in his hatred," Baldwin said, because he "prefers, like Lucifer, rather to rule in hell than serve in heaven."[17]

Here Faulkner was subtle: Joe might be ready to rejoin the human race, but that did not mean he wanted to be forgiven. Faulkner was ready for one of his boldest strokes. The charade that followed was stranger than anything Joe had yet experienced, for in Yoknapatawpha County martyrdom was not so easy to achieve. Joe might have foreseen what he was facing if he had remembered how easy it had been to elude the posse. "Like there is a rule to catch me by," he had reflected, "and to capture me . . . [any other] way would not be like the rule says" (*LIA*, 250). Now Joe walked boldly into a nearby town, but no one recognized him; to get arrested, he had to start a fight. And Wright's Max, who thought people's "sympathy" for killers like Joe was founded in "guilt," would have understood what happened when Joe was taken to Jefferson. A determined sheriff stood off the mobs. Gavin Stevens, Yoknapatawpha's aristocratic county attorney, offered Joe what Max had hoped for Bigger, a life sentence in exchange for a guilty plea. Former minister Gail Hightower was ready to alibi Joe by swearing they were involved in a homosexual encounter at the time of the murder. People like that, Faulkner wanted his readers to know, were as deeply involved in the rituals that produced lynchings as anyone else.[18] Some people wanted to kill Joe because of the guilt they projected onto him; others wanted to save him for the same reason, and they had done their work so well that Joe was finally faced with the strange task of orchestrating his own martyrdom. Joe was going to have to break *out* of jail in order to be lynched.

Once he had escaped, Joe was able to find people in Max's other category, people whose "fear" combined with their "guilt" to produce "rage." The man who now surfaced was Percy Grimm, a National Guardsman whose evangelistic chauvinism, like that of John McClendon in "Dry September," was the counterpart of Doc Hines's and the Burdens' Manichaean racism: his creed was "a sublime . . . faith in physical courage and blind obedience, and a belief that the white race is superior to any and all other races." What happened left people like County Attorney Stevens puzzled. Joe seemed about to get away; but Grimm, seizing a bicycle, found him almost instinctively and gave chase. Dodging into a black's isolated cabin, Joe found a pistol and seemed about to shoot Grimm with it; but instead, he circled to Hightower's house, bludgeoned the minister with the pistol, and barricaded himself behind the kitchen table. When Grimm charged in firing, Joe did not shoot back; and then, fatally wounded, he lay passively as Grimm achieved with a butcher knife what Joe had run from for "the thirty years that I have lived to make me what I chose to be." To the people of Jefferson, Faulkner reported, "It was as though he had . . . made his plans to passively commit suicide" (*LIA*, 336, 330); but Joe had done more than that, and the final irony of his martyrdom was how little such people understood.

County Attorney Stevens, the only aristocrat other than Hightower who was close to those events, sounded more than a little like Wright's Max as he struggled to cope with them. Max thought Bigger was a product of the "complex forces of society";[19] Stevens thought Joe had been shaped by similar forces. "There was too much running with him," he conjectured, for Joe to escape: "It was not alone all those thirty years" of Joe's life, but also the generations before him, "all those successions of thirty years" prior to his birth (*LIA*, 334). But there the likeness stopped. Max, Wright was trying to show, understood the forces that had produced Bigger; but the one man in Jefferson who had the best chance to understand Joe understood no more than anyone else. Max's explanation of Bigger's fate was deterministic; Stevens's explanation of Joe's was theological and racist. Joe Christmas, Stevens conjectured, was the victim of a war at the depths of his being between black and white strains in his blood; and, as he expounded that strange theory, it became clear that this aristocrat had been as thoroughly infected with Jefferson's pervading Manichaeanism as anybody else.

In Faulkner's Puritan Jefferson of 1932, the most liberal member of the community thought he was being generous when he said Joe's problem was ancestors who "had put that stain either on his white blood or his black blood, whichever you will." To Stevens, "his blood would not be quiet, let him save it," and it explained what looked like Joe's effort "to passively commit suicide." Joe's black blood had led him instinctively to the black's cabin, Stevens thought, "and then the white blood drove him out of there, as it was the black blood which snatched up the pistol and the white blood which would not let him fire it." In Stevens's mind that war between Joe's bloods was a cosmic struggle between the forces of darkness and light. It was Joe's "white blood . . . which rising in him for the last and final time" had brought Joe to Hightower's house, had "sent him . . . into the embrace of a chimera, a blind faith in something read in a printed Book." But Stevens was sure Joe's black blood had been too strong for absolution. Black blood was primitive blood, and it was regressive; it had "swept him up into . . . [an] ecstasy out of a black jungle" —a place "where life has already ceased before the heart stops," a place where "death is desire and fulfillment" (*LIA*, 334).

Stevens should have known better. Those conjectures were "a rationalization which Stevens made," Faulkner said later; what they showed about Joe was not that he was victim of a strange war in his blood, but that he was destined to be misunderstood, that "the people that destroyed him made rationalizations about what he was."[20] Joe's flight showed that Joe, at least, knew such conjectures were too easy. Bigger's death at the hands of those who hated him, Baldwin said, was "the only death which will allow him a kind of dignity or even, however horribly, a kind of beauty."[21] That was what Joe had been seeking. Joe had had no idea of who Grimm was, but he had been sure somebody like Grimm would surface, and the man had performed exactly as Joe had hoped. Ambushing Grimm from the cabin would have been easy; what Joe had wanted was for his pursuer to see the pistol, so that Grimm would have to shoot him. What was more, Joe had wanted the minister's house to be the scene of his martyrdom; that way he could die as the devil-nigger who had attacked a preacher. It had worked out better than Joe could have imagined. As Grimm entered, the fumbling Hightower had finally tried to alibi him. "Jesus Christ!" Grimm had sworn. "Has every preacher and old maid in Jefferson taken their pants down to the yellowbellied son of a bitch?" Joe was cornered; and Grimm's anger at Hightower's alibi sent him charging into the kitchen, ready to go to work with the butcher knife. "Now you'll let white women alone even in hell," he had cried (*LIA*, 345).

Bigger's crime, Max insisted, was "an act of *creation*"; Joe's death was an act of creation, and he had contrived to orchestrate it to the last detail. Joe had been running all his life from Puritans, like Grimm, who projected their guilt onto others and lynched this guilt in the name of God; now one of those men had allowed Joe to purge his own guilt, and Joe had managed that without any threat to the selfhood he had struggled all his life to preserve. Joe had found his black cross, and the irony was that this cross had given him his revenge as well as his freedom. As the "pent black blood" burst from his groin "like the rush of sparks from a rising rocket," Joe was free at last to seek the light: "upon that black blast the man seemed to rise soaring." But Joe's martyrdom, Grimm and his friends now knew, was no "ecstasy out of a black jungle." What jungle those men had seen was a very American one, and as they watched Joe's dying face they were suddenly aware of what they were seeing. The memory of that face was going to stay with them, "musing, quiet, steadfast, . . . not particularly threatful, but of itself alone serene, of itself alone triumphant" (*LIA*, 346). Joe was where he wanted to be, and the rituals those men were pursuing had been turned against them. Joe had spent a lifetime learning how to do that; he might not be "particularly threatful" now, but he had eluded them as they projected that guilt onto him, and they would have to continue to live with their guilt. Their victim was free, but they would live in bondage.

That, in Baldwin's terms, was the story of the " 'nigger': black, benighted, brutal, consumed with hatred as . . . [whites] are consumed with guilt," and Baldwin thought this story was woven so deeply into the fabric of American life that understanding it meant understanding the country. "Telling that story," he believed, was "inevitably and richly to become involved with the force of life and legend"; it showed "how each perpetually assumes the guise of the other, creating that dense, many-sided and shifting reality which is the world we live in and the world we make." That was how the "nigger" could free us. "To tell his story is to begin to liberate us from his image," Baldwin thought, and that made the whole process real: "it is, for the first time, to clothe this phantom with flesh and blood, to deepen, by our understanding of him and his relationship to us, our understanding of ourselves and of all men." But Baldwin had reservations about the way Wright had handled Bigger Thomas; the story of the "nigger," he feared, was "not the story which *Native Son* tells," and he had two complaints. The first had to do with Bigger's connection to the blacks among whom he was raised. Bigger's "significance" was "as the incarnation of a myth," but he had no real "significance as a social . . . unit." Bigger "has no discernible relationship," Baldwin argued, "to his own people, nor to any other people."[22]

Ex-Mississippian Wright and Mississippian Faulkner had a common problem. Making Joe "the incarnation of a myth"—Baldwin's "monster created by the American republic"—Faulkner had conceived the bold stroke of having Joe raised entirely by whites. That had allowed Faulkner to isolate those aspects of black character forced on blacks by Puritan society, and it had allowed him to see Joe without the aristocratic preconceptions that he had brought to Dilsey; the problem was that it had allowed Faulkner to avoid dealing with those preconceptions, too. Few blacks other than Joe appeared in *Light in August*; the only black given any characterization was the mammy who raised Gail Hightower, an ex-slave whose loyalty resembled Dilsey's. Joe might be a perceptive illustration of what white society did to blacks; but *Light in August* showed little about what blacks were like without white society. Wright's failure to show how blacks behaved when they were isolated from whites, Baldwin thought, meant that "a necessary dimension has been cut away": some illustration of "the relationship that Negroes bear to one another, the depth of involvement and unspoken recognition of shared experiences which creates a way of life."[23] That, fortunately, was not the necessity for *Light in August* that Baldwin thought it was for *Native Son*; but it showed how far Faulkner had to go as he searched for the realities of his South where blacks and Puritans struggled without the restraining hands of the aristocrat.

But Baldwin had a second complaint about *Native Son*: he objected to the weight of Wright's plot falling on "Max's long and bitter summing up." That speech was misleading, he thought; it encouraged people to continue in the very rituals Max claimed he was trying to stop. "It is addressed to those among us of good will," he wrote; the problem was that "it seems to say that, though there are whites and blacks among us who hate each other, we [the men of good will] will not." For Baldwin that was "a dream not at all dishonorable, but, nevertheless, a dream," and its consequences were ruinous. It assumed that "the black man, to become truly human and acceptable, must first become like us"—that is, a man of good will. But why should Wright's "monster created by the American republic" be a man of good will? Max's speech meant that "the Negro in America can only acquiesce in the obliteration of his own personality, . . . surrendering to those forces which reduce the person to anonymity." Baldwin concluded that "*Native Son* finds itself at length so trapped . . . by the American necessity to find the ray of hope that it cannot pursue its own implications."[24]

It was here that *Light in August* was rich. If Max's rationale was simplistic, its inadequacy convinced Baldwin that Wright could see no further; but Faulkner had shown how simplistic his man of good will could be. County Attorney Stevens was as deeply involved in those rituals that "reduce the person to anonymity" as anyone else, and his involvement revealed Joe's fate in all its complexity. Such was Faulkner's triumph in *Light in August.* Eight years before *Native Son,* and two decades before *Invisible Man* and *Go Tell It on the Mountain,* Faulkner had catalogued those rituals; in doing so, he had defined almost every avatar of the experiences that living in white Puritan society forced on blacks. Relieved of the burden of defending his own white class, Faulkner had awakened to a perceptiveness that mocked the mellow nostalgia he had felt for the world of his youth. Faulkner's new novel might have shown him little about how blacks identified themselves ethnically; but he had managed to tell the story of the "nigger" all the same, and in doing so he had achieved something like the freedom of which Baldwin later spoke. "To tell his story," Baldwin believed, "is to begin to liberate us from his image."[25] In telling this story, *Light in August* was almost infinitely suggestive; it would stand as Faulkner's most genuine achievement in his search for a South.

It left him at a promising stage of that search. *Light in August* seemed to show that, without paternalism, what happened between blacks and Puritan whites would continue to produce the kind of rituals into which Joe Christmas, Percy Grimm, Doc Hines, and the Burdens were locked; if that were true, Faulkner's novel presented a telling argument in favor of paternalism. On the other hand, although Faulkner had learned little about black ethnocentricity, his novel had opened up a world of potentials for understanding how white society determined some types of black behavior. Taken together, those two implications of *Light in August* ought to unlock secrets about the world of Faulkner's youth that had so far eluded him. If the Puritan South shaped black character through rituals like the ones described in *Light in August,* had the Old South, with its values of order and protectiveness, been any different? The answer to this question promised to tell Faulkner a great deal about the world of his youth. The question represented, in fact, a challenge greater than any he had yet risen to, for it was toward the Old South, not the South of Joe Christmas, that it pointed.

7

A Visit to a Dark House

I

In 1927, while *Sartoris* was still in manuscript, William Faulkner married recently divorced Estelle Oldham and assumed care of her children, Cho-Cho and Malcolm Franklin. A year later he purchased an antebellum mansion south of Oxford known, after its builder, as the old Shegog place; long in need of repair, it was positively Gothic by the time the Faulkners moved there. Young Cho-Cho and Malcolm had to get used to doing without electricity and plumbing for a while, but there were compensations. Their stepfather told them about Judith Shegog, whose ghost had never left her father's house; from time to time everybody in the family seemed to hear Judith playing Estelle's piano in the middle of the night, and one night Cho-Cho saw Judith in the bedroom.[1] Three years later the title Faulkner gave the first book he wrote in that mansion—he had named the place Rowan Oak—was *Dark House*. He began this manuscript with aristocratic Gail Hightower sitting beside his darkened window waiting for the ghost of his grandfather to appear out of the twilight.[2] Later, as Faulkner became preoccupied with the poor-white Lena Grove and "parchment-colored" Joe Christmas, he chose to emphasize the other pole in his Manichaean spectrum, and he renamed his novel *Light in August*; but Hightower, sitting alone in his decaying house, suggested a great deal about where Faulkner's search for a South was taking him.

Hightower stood for the first phase of that quest. Like Quentin Compson, this descendant of antebellum aristocrats seemed lost in a fantasy of a Cavalier past that he had built around his grandfather. But as a mesmerized Hightower watched the ghost of that old soldier emerging from the twilight, Faulkner must have realized that one important part of his struggle to find his South had ended in failure. His struggle had thus far revealed a bleak landscape. The values of order and protectiveness that he associated with the Old South seemed on the point of extinction; moribund among aristocrats like the Hightowers, Sartorises, and Compsons, they survived only in blacks like aging Dilsey Gibson. And what had taken their place seemed monstrous. There remained only a wasteland presided over by money-grubbing Flem Snopes and Manichaean racists like Doc Hines and Percy Grimm; the only southerners who retained their humanity were the Dilseys and Lena Groves, whose instincts seemed to save them in spite of the pervading decadence.

As Faulkner surveyed that wasteland South, he was faced with a question he could no longer avoid. He had shown how the decadence extended laterally, through both of the present-day white classes; could that mean that it also extended backward, permeating the South's past as well as its present? To this point, Faulkner had approached Colonel John Sartoris and Grandfather Compson the same way his family had approached Colonel W. C. Falkner: through the patchwork of fact and legend that his descendants had woven around their patriarch. But if the Old South had created a meaningful life-style, if a Simon Strother or a Dilsey Gibson had been nourished enough by that life-style to become attached to it, Faulkner ought to be able to show more graphically how that South had functioned. Rowan Oak stood for a new phase of Faulkner's search for a South. That phase, with which he would be preoccupied for the next decade, surfaced first in a tale on which he was working in the summer of 1931.

"Evangeline" was about long-dead Colonel Thomas Sutpen, whose house was in even worse condition than the Shegog place had been; his daughter Judith had lived in that house, and in its own way it was haunted, too. The story's anonymous narrator, a hackwriter, had learned about Sutpen through blacks in the neighborhood. They remembered that a man named Charles Bon had been killed, apparently by Sutpen's son Henry, with "the last shot fired in the war." When Henry fled and his father died, Judith had kept the plantation together through the Reconstruction, assisted by a mulatto woman named Raby. Judith was dead now, but Raby was still living in the house, taking care of Henry, who was home now and near death; as Faulkner's hackwriter questioned Raby, the old house seemed at last to be yielding its secret. "Henry Sutpen," she confessed, "is my brother"—then, as though to seal off her secret from the world, Raby burned her father's house to the ground with both of his surviving children in it.[3]

But in Sutpen's house there were secrets behind secrets; when one was peeled away, another revealed itself. The narrator thought he had found the house's real secret in its ashes: a metal case that Judith had given to Bon with her picture in it. It was not blonde Judith's countenance that looked out of that case, but something quite different: "the smooth, oval . . . face, the mouth rich, full, a little loose, the hot, slumbrous, secretive eyes, the inklike hair with its faint but unmistakable wiriness." For the hackwriter, the face had "all the ineradicable and tragic stamp of negro blood," and the picture was inscribed "*A mon mari*." Now, the hackwriter decided, he understood "what to a Henry Sutpen . . . would be worse than the marriage." Having a mulatto sister like Raby was one thing; having a white sister marry a man already married to a mulatto was yet another, and it "compounded the bigamy to where the pistol was not only justified, but inescapable."[4]

That was a species of shock Gothic enough to jolt the most heavy-handed abolitionist; but, earlier in the story, as the hackwriter speculated about the tragedy of Sutpen's house, a problem began to reveal itself. Perhaps "nowadays we can no longer understand people of that time," he mused. "Perhaps that's why to us their written and told doings have a quality fustian though courageous, gallant, yet a little absurd." Then, as though reflecting Faulkner's own uncertainties, he decided that "that wasn't it either." Raby had been holding back "something more than just the relationship between Charles and the woman; something . . . which I knew she would not tell out of some sense of honor or of pride"; and he concluded, almost as though Faulkner were talking to himself, "And now I'll never know that. And without it, the whole tale will be pointless, and so I am wasting my time."[5]

"Evangeline" was no ordinary story. Just when Raby's secret seemed about to emerge, it developed that she was hiding another, deeper secret; and before that secret was resolved the narrator stepped in and hinted at still another, a secret he did not quite understand either. Faulkner was dealing with material that was difficult for him. To this point he had looked at the failures of the aristocratic tradition only in the modern descendants of the Cavalier families; "Evangeline" called for a look at the failures of the Cavalier patriarchs themselves, and something had happened to Faulkner—he ended up complaining that his own tale was "pointless," that he was "wasting my time." Yet Sutpen's story had taken hold of him, and when he tackled another phase of it in 1933 he was clearly ready to "waste" more of his time if necessary.[6] "Wash" was about a redneck named Wash Jones who had put his faith in Sutpen in somewhat the same way old Falls had put his in John Sartoris in *Sartoris*. "I know that whatever you handle or tech," Wash told his patron, "whether hit's a regiment of men or a ignorant gal or just a hound dog, that you will make hit right" (*CS*, 542). That "ignorant gal" was Wash's granddaughter Milly, and she was pregnant by Sutpen, who wanted a male heir to take Henry's place. When Milly gave him a girl instead, Faulkner unfolded a new aspect of Sutpen's personality: the old man seemed more interested in one of his mares that had foaled. Colonel John Sartoris died in an affair of honor; but there was no honor in what happened to Sutpen. Wash killed him with a scythe, and this poor white's anguished cry sounded the tragic gap between Mississippi's two white classes: "Better if his kind and mine too had never drawn the breath of life" (*CS*, 548).

For Colonel W. C. Falkner's great-grandson, that was a new note, and it gave another dimension to the abolitionist version of the plantation that had emerged with "Evangeline." In the Sutpen story Faulkner was ready to fuse the complaints of both of America's Puritan classes—the northern abolitionist and the southern poor white—into one major novel that would explore the Old Order's failures. "Roughly," he wrote publisher Hal Smith, "the theme is a man who outraged the land, and [who found that] the land then turned and destroyed the man's family."[7] Faulkner was going to give his new novel the name he had originally planned for *Light in August,* the first novel begun at Rowan Oak: he was going to call it *A Dark House.* Smith could tell how serious his writer was by a decision Faulkner had made: he was going to ressurect Quentin Compson for his "protagonist" to replace the hackwriter-narrator of "Evangeline."

One lengthy work should have provided ample analysis of that young man's problems; if Faulkner was bringing Quentin back, it was an acknowledgment that he was still preoccupied with Quentin's tragedy, that *The Sound and the Fury* had left some issues unresolved.[8] Smith could hardly have guessed that from the reason Faulkner gave. Quentin would "keep the hoop skirts and plug hats out," Faulkner said, and that would save *A Dark House* from being just another "historical novel"; but Faulkner's comments to Smith also suggested something else. In *The Sound and the Fury* Quentin had killed himself because he chose not to face a world in which the values of his aristocratic grandfather, which he had projected onto his sister Caddy, were challenged; now, Faulkner indicated, Quentin's feelings about his homeland were mixed. "I use him," he wrote Smith, "because it is just before he is to commit suicide because of his sister, and I use his bitterness which he has projected on the South in the form of hatred of it and its people."[9] *The Sound and the Fury* had revealed a twentieth-century paternalism that had lost touch with its nineteenth-century values; *A Dark House* would suggest that the values Quentin thought his family had once lived by might never have existed.

When someone later asked what kind of trouble the novel had given him, Faulkner remembered that it had been "very difficult." He had "worked on that for a year" and then had to "put it away." He was still not sure what had happened. "I think that what I put down [first] were inchoate fragments that wouldn't coalesce," he said. "I decided that I didn't know enough at that time maybe, or my feeling toward it wasn't passionate enough or pure enough." Whatever the problem, he had "got in trouble" with that novel, and he had made a major decision: "I had to get away from it for a while, so I thought a good way . . . was to write another book." That, Faulkner said, was why he had written *Pylon* (1935).[10]

Pylon, with its twentieth-century urban setting, seemed like a good opportunity "to get away" from the Old Order's failures; but, in fact, Faulkner had written two books between the time he began to bring "Evangeline" and "Wash" together as *A Dark House* in 1934 and the time he published the work as *Absalom, Absalom!* in 1936. Pressed for money, he began a series of stories for the *Saturday Evening Post*; "An Odor of Verbena," written just before he brought this series of stories together as *The Unvanquished* in 1938, was the only one of those tales not written while Faulkner was struggling to "coalesce" those "inchoate fragments" of *Absalom, Absalom!*[11] The stories amounted to a complete novel without "An Odor of Verbena," and they provided a much better way "to get away" from the Old Order's failures than did *Pylon.*

If Faulkner's feelings about that Gothic South he had been exploring in *Absalom, Absalom!* were not "passionate enough or pure enough," it was no wonder: in this new series Faulkner was back with the Sartoris family, surrounded by the comfortable myths of his youth. "I had never thought of . . . [that material] in terms of a novel," he said later, only "as a series of stories." But one led to another, and soon "I had postulated too many questions that I had to answer for my own satisfaction," and other stories "had to be."[12] It was one of the strangest episodes of Faulkner's search for a South. He had set out to write about the Old Order's failures; he had produced instead another mellow defense of its virtues.

The distance between *Absalom, Absalom!* and this new Sartoris series was clear from Faulkner's choice of a narrator. Colonel John's son Bayard—the old Bayard of *Sartoris*—was of a different order of man from Quentin. Bayard had been twelve when Vicksburg fell in July 1863, so he was almost the same age fourteen-year-old J. W. T. Falkner had been. As Bayard began his narrative, it sounded like a fantasy of how Faulkner's grandfather might have learned what being an aristocrat meant while Colonel W. C. Falkner was away fighting.[13] Colonel John was almost a mythical figure for his son; and, as the fighting came closer and a maturing Bayard struggled to understand its meaning, Faulkner turned to family memories that had been on his mind since *Sartoris.*

Bayard's mother was dead; but there was something of Auntee Holland Falkner Wilkins in Rosa Millard, the grandmother who ran things in Colonel John's absence, and something of Damuddy Lelia Swift Butler, too. As "Granny" emerged as Bayard's model of paternalistic responsibility, the meaning of those "questions" Faulkner had "had to answer for my own satisfaction" became more obvious. *Sartoris* had been a sentimental farewell to the world of Faulkner's youth; this new series was intended to explain the events that had produced that world.

Miss Rosa was a subtle choice. A devout Christian, she was as committed to the female aristocrat's role as guardian of propriety as Damuddy Butler had been; but in wartime people's roles were sometimes reversed. Granny was no Manichaean Puritan; she was an Episcopalian, as committed to preserving the aristocracy as Auntee Wilkins, and with Colonel John gone she had to take over the male aristocrat's role of protector. Granny's commitment was something new for Faulkner. He had made no apology for the Sartoris family's status in *Sartoris;* but now, unable to finish his tale of how Thomas Sutpen had abused that status, he was implying something different about the Sartorises. Glamorous Colonel John, Miss Rosa's portrait implied, was not merely interested in aggrandizing himself and his family, as Sutpen was; he was fighting because, like Miss Rosa, he felt responsible not only for the Sartorises but also for the entire countryside.

As the war dragged to its close, Granny became Faulkner's tragicomic symbol of that commitment. At first this formidable matron seemed capable of sustaining both male and female roles. When Bayard and his black friend Ringo impetuously fired on a Union officer, she enveloped black and white boy alike in her billowing skirts while the furious Yankees ransacked the house; then she made both boys pray with her because she had lied for them. But changing roles was tricky. What the war was doing to Granny surfaced when the Union troops took the livestock and silver and burned the Sartoris place: now Granny joined Bayard and Ringo shouting "bastuds" as those Yankees without apology.[14] That incident was the first sign that the war was tilting Granny's commitment permanently out of balance; the proof came when she went to a Union colonel to demand her property back and a careless quartermaster gave Granny more mules and silver than she had lost. If Granny accepted this deceit—against the very enemy who had destroyed her home—she could keep the family and its slaves going.

That, Bayard learned, was the war's real irony: Granny's Episcopalian conscience was its first casualty. This devout lady now became as accomplished a mule smuggler as any scalawag, distributing her booty to hillmen and free blacks to work their land, even selling some back to the

Union troops for money, which she gave to charity. Now, as Granny prayed, there was a certain arrogance in her manner. "I have sinned," she admitted to God, but she was sure her motives were pure. "I did not sin for gain or for greed," she insisted, and as Granny recited her sins they sounded more like a catalog of paternalistic virtues. "I sinned first for justice," she told God, but Granny wanted God to know that later she had been sinning "for more than justice": "I sinned for the sake of food and clothes for Your own creatures who could not help themselves." By the time Granny told God that "if this be sin in Your sight, I take this on my conscience too,"[15] it was clear she was developing a kind of benign hamartia of her own. It came as no surprise when Miss Rosa got in over her head and was murdered by the scalawag Grumby.

The first phase of Bayard's education ended with a bitter lesson: it was this loyal soldier of the home front, not reckless John Sartoris, who had been the first to fall defending the Old Order. But Bayard's education was not complete. Granny had given her life to show Bayard what being responsible meant; but there was another side to Bayard's heritage, and, here, with Colonel John still off fighting, Bayard had to be his own mentor. "I want to borrow a pistol," he told aging family friend Buck McCaslin. At fifteen, with only Ringo to help him, Bayard was ready for a gesture a Cavalier's son might make. When John Sartoris returned, Uncle Buck took him to the old compress where Granny had been murdered, and "the first thing we see was that murdering scoundrel [Grumby] pegged out on the door." Uncle Buck, at least, knew what else to expect: Grumby's right hand was missing. "And if anybody wants to see that too," he said, "just let them ride into Jefferson and look on Rosa Millard's grave!" For Uncle Buck, that proved something. "Ain't I told you he is John Sartoris' boy?"[16]

John Sartoris's boy had learned paternalism from a woman, and he had taught himself what honor meant; but his education was still not complete. The abolitionist Burden family was trying to elect a black sheriff that Jefferson's whites regarded as a corrupt clown. Bayard, separated momentarily from his father, reached town just in time to see Colonel John's old troopers cordoning off the Burdens' hotel and to hear three shots. "I let them fire first," his father said when he emerged, but it was only when he refused the bodyguard his old soldiers offered that Bayard realized the degree of his father's commitment. "Don't you see we are working for peace through law and order?" Colonel John said. That was why he had forced the Burdens to fire first: he meant to be tried and acquitted. "I will make bond," he said.[17] The Redemption had come early to Jefferson, and if there was a certain contradiction in Colonel John's notions of law and order, at least he was spelling things out: Jefferson might have been ravaged, but the Old Order in the person of John Sartoris was back to stay.

That was where Faulkner rang down the curtain on his tragicomedy—with John Sartoris running the community and an emerging Bayard ready to step in. It was not until 1938, three years later, that Faulkner added a final tale, "An Odor of Verbena." Those Sartoris stories had marked a strange interlude in Faulkner's search for a South. In *Sartoris* Faulkner's affection for his fictional family, so closely modeled on his own, had been clear, but he had made no apologies for their social role; now, stalled in his efforts to visit the Gothic South of Thomas Sutpen, Faulkner had stumbled onto a new quality of sentimentality. These new Sartoris stories were more than an affectionate reminiscence; they were a parrotlike re-creation of the Old Order's official rationale, as blandly propagandistic as any story written by Thomas Nelson Page. What had happened became obvious as soon as Faulkner turned to the Sartorises' blacks. When Faulkner had first begun to tell the story of the world of his youth in *Sartoris,* his thoughts had turned to a family of blacks who like Caroline Barr and Ned Barnett had served the Falkners: to "one-armed Joby" and "old Louvinia who remarked when the stars 'fell.' " In his new series those figures from his youth were on his mind again. As in that earlier book, Joby Strother was the wizened retainer who had come with the Sartorises from Carolina; his son Simon, the aging groom of the earlier novel, had followed John Sartoris to war. Now Faulkner added new members to Joby's clan. Louvinia, a bustling, practical-minded mammy, was Joby's wife; she had a son named Loosh, who was Simon's brother; and Bayard's companion Ringo was Simon's son.[18]

As Union troops neared Jefferson, Loosh would break out nights to bring back news. Soon he was acting like a man possessed; with a "look on his face . . . like he had not slept in a long time and he didn't want to sleep," Loosh told anyone who would listen that "Gin'ral Sherman gonter sweep the earth and the race gonter all be free!" By the time the troops finally arrived he had elevated his freedom to theology. "God's own angel," he declared, "proclamated me free and gonter general me to Jordan." For Bayard, Loosh's reaction showed what Colonel John was fighting for: anybody deserved freedom, but Loosh was incapable of coping with it. "I don't belong to John Sartoris now," he told Granny. "I belongs to me and God"—and he was ready to reward his rescuers by showing them where the family silver was hidden. "But the silver belongs to John Sartoris," Granny objected. "Who are you to give it away?" Loosh was so far gone that he could not distinguish himself from the other property Colonel John had owned. "Let God ax John Sartoris who the man name that gave me to him," he told her. "Let the man that buried me in the black dark ax that of the man what dug me free."[19]

Loosh's rebellion showed the distance those Sartoris stories had taken Faulkner. Joe Christmas's revolt against McEachern, despite its Faustian quality, had come through as a somehow admirable, even heroic act; in Thomas Sutpen Faulkner had already conceived the notion of "a man who outraged the land, and [who found that] the land then turned and destroyed . . . [his] family." In an Edenic sense Loosh was right: John Sartoris, like Sutpen, had no more right to own the earth's goods than he had to own its people; but for the moment Faulkner seemed unable to admit that the worlds of Joe Christmas and Thomas Sutpen could overlap that of John Sartoris. Loosh, at any rate, was no Joe Christmas, daring existential freedom as devil-nigger; he resembled, rather, the bumbling Caspey of *Sartoris,* with his fantasy that World War I had "unloosed de black man's mouf."

Loosh's next response to his freedom showed again why John Sartoris had to be in charge of it, and the silver, too: he joined one of the bands of blacks who were roaming the countryside in search of the River Jordan, which they were convinced was nearby. Drusilla Hawk had seen the spectacle. "We couldn't count them," she told Bayard, "men and women carrying children who couldn't walk," some "carrying old men and women who should have been at home waiting to die." To Drusilla it was as though they had thrown themselves into some primitive trance. "They were singing," she said, "walking along the road singing, not even looking to either side," and they had lost all touch with reality. "The next morning every few yards along the road would be the old ones who couldn't keep up any more, . . . calling to the others to help them: and the others . . . not stopping or even looking at them." Drusilla was sure something terrible was about to happen. " 'Going to Jordan,' they told me."[20] When that motley cavalcade accidentally came upon a river, they swarmed toward an unfinished bridge; to prevent catastrophe Union troops dynamited it, but by then many of the marchers were drowned or trampled. That incident, Faulkner seemed to be suggesting, spoke volumes about black character. There was a tiny minority of adults (of whatever age) like Ringo and Louvinia who had sense to stay with their masters; and there was a vast majority of simple-minded children ready to strike out for an impossible freedom. But if Faulkner's novel threatened at this point to founder in a mire of paternalistic propaganda, it was here, paradoxically, that he was also at his most subtle. For between his readers and that crude stacking of the deck, he was ready to interpose one of his shrewdest creations, Ringo Strother.

Again Faulkner turned to familiar sources. Aging Ned Barnett, John Faulkner wrote, had been "a slave belonging to my great-grandfather"; a younger man than J. W. T. Falkner, Ned had probably been born toward the end of the war.[21] But the quality of the relationship between Ned and the Young Colonel posed a problem that, perhaps, raised one of those "questions" about the Sartoris family Faulkner "had to answer for my own satisfaction": what would that relationship have been if Ned had been the same age as the Young Colonel? Ringo, a youthful version of that crochety old man, was as secure in his status as Ned; a third-generation Sartoris servant, Ringo had been raised like a brother to Bayard. Theoretically Ringo spent his nights on a pallet in Bayard's room; actually both boys slept in the bed like brothers, and they were so close Ringo called Miss Rosa "Granny," too. Under her pious supervision they reached puberty as boon companions, competing amicably for Colonel John's attention during his rare visits. Ringo, ironically, matured ahead of Bayard, and his achievements were intellectual as well as physical. A precocious artist, Ringo forged Union requisitions with alacrity, and he outstripped Bayard as Granny's assistant in the mule-smuggling business; soon Granny was trusting him with the accounts and sending him on complicated missions; when Granny was killed, Ringo turned out to be Bayard's mainstay in tracking down Grumby. The Sartorises, it seemed, were so open with trusted blacks, and Ringo so able, that this youth had become indispensable to his white family; and Ringo, caught up in his role, had begun to identify himself wholly with their ethic.

He became indispensable to Faulkner, too. Ringo was the one ingredient necessary to make those stories work. Giving Bayard the role of Granny's chief assistant would have committed all of the Sartorises to the benign hubris that killed her; but, more important, Ringo was the perfect foil for Faulkner's robotlike Jordan marchers. They had shown Bayard how most blacks could not survive their freedom; Ringo taught him how one aggressive, intelligent slave could be truly free. The only trouble with Ringo was his virtuosity, which sometimes bordered on the incredible; but Faulkner knew how to deal with that, too. Whenever things seemed to be getting out of hand, Faulkner allowed another side of Ringo to emerge. Ringo, like Ned Barnett, was garrulous, and his talk exuded a humor that masked his precocity. At first the jokes seemed to be on Ringo. "Vicksburg fell?" twelve-year-old Ringo asked when he heard the city had surrendered. "Do . . . [that] mean hit fell off in the River?" Soon he was using his own wit to probe other people's foibles. He saw through Granny's rationalization that "the hand of God" had provided her with those mules: "Whose hand was that?" he asked when he wheedled twelve horses out of the Yankees. Ringo's wit defined his social position. "That's one more mouth to feed we got shed of," he told Granny when she sent a stray black to the Union troops. And his predictably ironic response to emancipation was "I aint a nigger any more. I done been abolished."[22]

If all that sounded familiar, it was. Ringo's crotchety aggressiveness might resemble that of a Ned Barnett, but Faulkner was drawing on sources older than Ned. In their own way the slave who refused freedom and the loyal retainer who helped run the plantation while his master was fighting Yankees had served the Old Order's nineteenth-century defense as ably as Jeb Stuart or Stonewall Jackson, and only the art that held such threadbare traditions together in Ringo was new. The trouble with Ringo ran deeper, though. There was probably no denying that the Old Order had seemed attractive for some blacks; but Ringo, like Dilsey Gibson and Simon Strother, stood only for that small number of blacks favored as members of white households. This youth, who seldom fraternized with field hands, suggested little about the realities of black experience; the wartime acceptance he had achieved was virtually unique, and it would not last. By the time Ringo and Bayard had lashed Grumby's hand to Rosa Millard's grave, what a mature Ringo was going to face should have been obvious. If Ringo had had anything to do with that grisly triumph, one would never have known it from the way Uncle Buck was chortling about Bayard: "Aint I told you he is John Sartoris' boy?" he kept repeating. "Hey? Aint I told you?" John Sartoris was back now. No precocious black boy was needed as Colonel John got the Redemption started in "Skirmish at Sartoris," and the difference between Ringo and Bayard was about to emerge. To Ringo, it should have been clear that identifying himself with John Sartoris was as much of a delusion as running around the countryside looking for the River Jordan. Southern society produced no black Cavaliers.[23]

If Ringo had understood his predicament, it was not obvious from Bayard's narrative. Ringo, like Granny, stood for that wartime situation in which everybody's role was reversed; examining this new role would have undermined the smug re-ordering of southern life in "Skirmish at Sartoris," and Faulkner could do no more than ease Ringo into the background. What was more, if Ringo provided the solution to some of Faulkner's questions, he also posed many new questions, and by the time Faulkner had finished "Skirmish at Sartoris," he knew a great deal about where those new questions were taking him.

Ringo, who was Simon's son, had not appeared in *Sartoris;* instead, Faulkner had given Simon a daughter, Elnora. This "tall, coffee-colored woman" (*CS,* 728) had little to do in that novel, but Faulkner brought her back in "There Was a Queen" in 1932, a story set ten years after the death of young Bayard, grandson of Faulkner's narrator in the new Sartoris series. Now only Bayard's widow Narcissa, their son Benbow, and ninety-year-old Virginia Du Pre remained from the Sartorises. Elnora seemed to be holding the surviving Sartorises together much as Dilsey had held together the Compsons; but she was quite literally a different breed from Dilsey, and, as she watched Narcissa with an "expression of quiet and grave contempt," another reason for her loyalties emerged. "Town trash," Elnora told herself, and she ruminated on what she would like to tell this intruder. "We never needed you," she said voicelessly (*CS*, 728, 729).

The Sartorises' maid, like Dilsey, had never doubted that she was a black member of the family; but Faulkner was ready to explore a new irony in this family's tragedy. Elnora, he declared, was John Sartoris's daughter, old Bayard, "her half-brother." For Faulkner that was a startling admission that threw the whole structure of myths that he had been building around the Sartoris and Compson families in a new light. It gave meaning to Elnora's status as black member of the family that had not been visible in Dilsey's situation; and it gave a genuinely convincing reason for her loyalty. And, if Colonel John turned out to be no more than human, a new dimension was added to the mellow legends that had accumulated around him. It also brought a new irony to the deaths of the Sartoris men: young Benbow, it appeared, was going to be ruined by Narcissa, and thus the Sartoris tradition would remain in Yoknapatawpha County only in the memories of their blacks. And what followed was even more startling. In a queer, ungrammatical aside, as though Faulkner himself were unwilling to admit that Colonel John was raising a white son and a black daughter in the same house, he added that "possibly but not probably neither of them knew it, including Bayard's father" (*CS*, 728, 727).

The Sartorises were sacrosanct, it appeared; they had to be approached delicately, even by their creator. If Bayard, at least, knew anything about Elnora and the circumstances of her birth, it was not evident as he narrated the stories that became *The Unvanquished*. Faulkner had good reason to keep quiet about this family skeleton; if Ringo was Elnora's brother, he could have been Bayard's, too. Still, it must have been clear to anyone who had read "There Was a Queen" why Faulkner was having trouble with Ringo: another novel was threatening to emerge from beneath the smug nostalgia of Bayard's narrative. That new novel was about two boys who really were brothers. The mother of the white brother had died, but the woman who had taken her place was not the mother of the dark brother; she was the white brother's grandmother—who insisted on acting not only as his mother but also as the mother of the dark brother. As these boys competed amicably for their father's attention, neither seemed conscious of their tangled family relationships —or that one of them was white, one black. But as they matured, their pigmentation became more important than any other factor in their lives, and their casually accepted values became reversed; the dark young man realized that his beloved brother and revered father were his enemies. This was the deeper, more tragic story which underlay Faulkner's new Sartoris tales, and by the time he had finished "Skirmish at Sartoris," Faulkner was aware of it.

He had reached a turning point in his search for a South. Behind the bland mask of John Sartoris, the secret of Sutpen's dark house had been laid bare. At last Faulkner knew what Sutpen's daughter Raby—whom he now renamed Clytie—"would not tell out of some sense of honor and pride." There was no longer any reason to feel that the tale was "pointless," that "nowadays we can no longer understand people of that time." Faulkner had discovered that Raby/Clytie was going to preside over the struggles of his middle years, a ghost more threatful than any that he had inherited with Rowan Oak.

8

Clytie's Secret

I

Rosa Coldfield, Thomas Sutpen's sister-in-law, had spent a long life struggling to untangle the secret that had confounded the hackwriter-narrator of "Evangeline." Hurrying to Sutpen's Hundred that day in 1865 when Henry killed Bon, Miss Rosa found Clytie blocking the stairs to Judith's bedroom. "Dont you go up there, Rosa," Clytie told her, and Miss Rosa, who had no way of knowing what secret Henry was explaining to her niece, was traumatized. "My entire being," she told Quentin, "seemed to run at blind full tilt into something monstrous and immobile." What shocked Rosa was Clytie's "coffee-colored face"; it stood for something she could not quite name. It was an African face; it exuded "a brooding awareness . . . of the inexplicable unseen, inherited from an older and a purer race than mine." But it was also a "Sutpen face"; it was like a "replica of his own which he had . . . decreed to preside in his absence." No wonder that, when Clytie spoke those words, Miss Rosa felt "as though it had not been she who spoke but the house itself." Miss Rosa should have been able to guess what Henry was telling Judith; the secret was written all over Clytie, who had come to stand for everything that Puritan Miss Rosa feared about the man her sister had married. In this "coffee-colored" little woman Sutpen had "created in his own image the . . . Cerberus of his private hell" (*AA*, 138, 139, 136).

Now, years later, Miss Rosa was trying to show Quentin Compson her vision of the curse that Nathaniel Burden had warned his daughter about, and this vision was going to be even more traumatic for Quentin than it had been for Puritans like herself and Joanna. Quentin had spent most of his short life trying to keep from seeing behind the aristocratic masks he had projected onto Caddy and his grandfather; in *The Sound and the Fury* the thought that there might be another reality behind those masks had driven him to suicide. Now Faulkner had brought his suicidal hero back to register the full impact of this vision, and the decision showed the course his career was taking. One lengthy book should have provided ample analysis of this young man's motives; that Faulkner was bringing him back was a kind of admission that he had not satisfied himself about Quentin. But if Faulkner was going to break through to the realities behind the masks Quentin was projecting onto Caddy and General Compson, it was time for Faulkner to identify these motives. His struggle to preserve the world that, before *Sartoris,* he had been "preparing to lose and regret" had been going on for nine years, and he was reaching a crossroads.

In the first phase of Faulkner's search for a South he had unveiled the twentieth-century South of *Sartoris, The Sound and the Fury,* and *Light in August*: a wasteland where Cavaliers like General Compson and Colonel John Sartoris were no more than legends, a South where blacks and Puritan rednecks struggled without the responsible hand of the aristocrat to guide them. In the second phase of his search Faulkner had explored those legends, and he had uncovered two older Souths: the South of John Sartoris and the South of Clytie's secret. But the more Faulkner explored those older Souths, the less they seemed to resemble each other. In Faulkner's *Unvanquished* stories dashing John Sartoris, with his aristocratic pedigree and his paternalistic sentiments, had begun to sound like little more than another, more subtle incantation of the paternalistic propaganda of the world of Faulkner's youth. These stories had come easily for Faulkner. But Clytie's secret pointed in the opposite direction, suggesting the tragedies of slavery and miscegenation that Puritan Yankees and Puritan rednecks decried. It was here that the writing had been difficult; Faulkner had found himself working with "inchoate fragments that wouldn't coalesce" and had "got in trouble." That was why it had taken Faulkner three years to complete *Absalom, Absalom!* and it was why Quentin, struggling to come to grips with his fears about his family's past, had to be its narrator. If his "protagonist" was projecting "bitterness . . . on the South in the form of hatred of it and its people," the reason was to be found in Sutpen's dark house.

For this new round with Quentin, Faulkner turned to a tradition of the southern novel as old as John Pendleton Kennedy's *Swallow Barn* (1832): a dialogue between a southerner and an outlander over the nature of southern reality.[1] The setting, appropriately distant from Mississippi, was Quentin's Harvard dormitory room, and for the other voice in his dialogue Faulkner chose Quentin's Canadian roommate, Shreve McCannon. Shreve, who ought to have been free from the usual Yankee prejudices, wanted to learn about Quentin's homeland. "You cant understand it," Quentin responded. "You would have to be born there" (*AA,*

361). But as Quentin struggled to piece together what he had learned about Sutpen from Miss Rosa, from his father, and from his own experiences, Shreve was soon complaining that Quentin did not know the South, either. There was more than one Sutpen, it seemed, and there were as many Souths as there were Sutpens.

"He wasn't a gentleman," Miss Rosa insisted, but, as she tried to explain what Sutpen was, he sounded like a Puritan's idea of a Cavalier. "I'll learn you to hate two things," Nathaniel Burden's father had told him in *Light in August*, "hell and slaveholders"—and Nathaniel, convinced slaveholders were in league with Satan, had decided that "the curse of the white race is the black man." Miss Rosa came from less belligerent stock than the Burdens, but the Manichaean framework of her Methodist faith was similar to theirs, and she projected the same demonic mask onto Sutpen that the Burdens projected onto slaveholders. Sutpen, Miss Rosa was sure, had brought a "curse on our family," and as she struggled to explain that curse to an amused Quentin her imagination invoked an "ogre-shape."[2] "Out of a quiet thunderclap he would abrupt," Quentin remembered, the "faint sulphur-reek still in hair clothes and beard." And for Miss Rosa, as for the Burdens, the Manichaean signature of that curse was Sutpen's slaves; in that picture Quentin always saw "behind him his band of wild niggers like beasts half tamed to walk upright like men" (*AA*, 14, 21, 13, 8).

Miss Rosa's devil-Cavalier sounded larger than life; the Sutpen whom Quentin's aristocratic father pictured was an upwardly mobile redneck and seemed smaller. "Sutpen's trouble," Quentin told Shreve, "was innocence," and, as he pieced together what his father had told him about the quality of that innocence, it took on a Manichaean tone that made Sutpen sound more like the Burdens than like Miss Rosa's ogre. Sutpen had been born in the country now known as Appalachia, in the mountain areas that later became West Virginia. There, Mr. Compson said, people like the Sutpens thought that "the land belonged to anybody and everybody" and that "the man who would . . . fence off a piece of it and say 'This is mine' was crazy." But that Edenic state had been lost to Sutpen when his father moved the family to the Virginia Tidewater. "They fell into it," Quentin mused, and as he described that Fall it sounded like a parable of a Mississippi hillman discovering the Delta. There were no blacks in the mountains, but now youthful Sutpen saw them in swarms, and he was soon projecting the same demonic attitudes onto them that Miss Rosa would later project onto him (*AA*, 220, 221, 222).

Those projections began in social outrage. Many blacks dressed better and had better houses than the Sutpens, and Sutpen found there was no way to channel his rage. Those differences existed not because there were differences between blacks and whites but because "there was a difference between white men and white men" (*AA*, 226). A person other than Sutpen might have sloughed off his Manichaean innocence and lapsed into an easy tolerance of class differences, but something happened that made tolerance impossible. On an errand to a Tidewater mansion, the boy was met by a liveried black who told him to go to the back door, and suddenly he found himself "seeing his own father and sisters and brothers as . . . the rich man . . . must have been seeing them all the time"; to that planter they were no more than "cattle" (*AA*, 235). Sutpen was assuming a lot for his aristocrat; in his innocence he believed everybody saw the world in Manichaean terms. But a curious thing was happening to Sutpen because of that assumption. Puritans like the Burdens, Mr. Compson's narrative seemed to imply, preserved their "innocence" by making slaveholders the representatives of darkness and casting themselves as defenders of light; Sutpen was about to preserve his through a similar Manichaean suspension, but the difference meant everything. Spawned in his mountain "innocence," Sutpen's Manichaean projections seemed almost instinctual; they had never been shaped by Puritan social tradition, and they produced a very un-Puritan reaction in Sutpen. Assuming both the role of light and the role he assigned to the planter, Mr. Compson's Sutpen appeared to project his notions of darkness onto everything that might stand in the way of his rise to affluence.

That, Mr. Compson conjectured, was when Sutpen had undertaken what he later called his "design": he was going to own a plantation someday, and no son of his was ever going to be turned away from anybody's door. "All of a sudden," Quentin said, "he discovered, not what he wanted to do but what he just had to do," and his future began to fall into line. "You got to have land and niggers and a fine house," Sutpen decided, "to combat them with" (*AA*, 263, 220, 238). By 1833 an older Sutpen had acquired a hundred-acre tract in Yoknapatawpha County from Chickasaw Chieftain Ikkemotubbe. Naming it "Sutpen's Hundred," he cleared it with twenty slaves he had mysteriously acquired and built a magnificent mansion; then he achieved the status he desired by marrying Miss Rosa's sister Ellen, an impeccable Puritan woman who gave Sutpen the children, Judith and Henry, who were to inherit the fruits of his "design." That achievement had been blasted with "the last shot fired in the war," when Henry killed Judith's fiancé, Charles Bon, and fled into exile.

Mr. Compson's redneck Sutpen was more credible—and more human—than Miss Rosa's devil-planter; but, more important, Mr. Compson thought he had uncovered Clytie's secret. Sutpen, he told Quentin, had discovered that Bon was secretly married to an octoroon in New Orleans through an illegal morganatic ceremony. He had forbidden the marriage, and, hoping Henry would help forestall a confrontation with Judith and Ellen, he had told the boy. But Henry and Bon, already fast friends, had gone off first to New Orleans, then to the Army of Virginia, thus giving fate the chance to make a decision neither was willing to make himself; it was only when both had survived that Henry, to prevent what he considered a tainted, bigamous marriage, killed his friend. With that, Mr. Compson believed, Sutpen's Manichaean "innocence" had carried him full cycle. His "design," spawned in that innocence, had begun when he had decided that no son of his would be turned away from anybody's door; now Sutpen's son had turned someone else away, and it had cost him his "design."[3]

Was that Clytie's secret? Mr. Compson had a mind for facts, and he could explain things Miss Rosa could account for only as the workings of an ogre; but for Quentin there was something missing in his father's solution. Why, if Sutpen had worked all his life to consummate his design, had he allowed something as unimportant as an illegal marriage to stop it? Clytie, by her very presence, showed how insignificant a morganatic marriage was in the eyes of all of the Sutpens. Mr. Compson was beginning to sound like Faulkner's narrator in "Evangeline" who thought that "nowadays we can no longer understand people of that time." Mr. Compson's problem was Bon. Faulkner's "Evangeline" narrator had learned that Raby/Clytie was Henry Sutpen's sister and that Bon was married to a black woman; Mr. Compson had learned all these facts, too, but he could not seem to get past them. Locked into his aristocratic sensibilities, he was exactly where Faulkner had been in the *Unvanquished* stories with Ringo: the clues he needed to lay bare the story of the Sutpens' dark brother were all around him, but his perceptions would take him no further. "It's just incredible," he told Quentin at one point. "It just does not explain" (*AA*, 100).

Faulkner was saving the full revelation of Clytie's secret for his "protagonist": it remained for Quentin to guess the truth, so that he and Shreve were able to put together an even more complex Sutpen. Quentin's father, it appeared, had passed on, without understanding, certain facts about Sutpen that he had learned from General Compson. Sutpen's Hundred represented Sutpen's second, not his first, attempt to establish that "design." The first was in Haiti, and he seemed to have succeeded when he saved the lives of his employer and his family by putting down a black rebellion. In telling the story Sutpen made himself sound like a hero in a cosmic Manichaean struggle: he simply "went out and subdued them," he told General Compson. His reward was marriage to his employer's daughter. But his triumph would not be quite so simple. Calvin Burden, who thought his grandson had a "black dam" and a "black look," could have told Sutpen that the more one wrestled with the dark powers, the more they left their mark. Sutpen discovered this when his wife gave him a son. For the first time he became aware of a fact that would "have . . . voided and frustrated . . . the central motivation of his entire design": his wife had African blood. This was the reality that had driven Sutpen from Haiti to Mississippi, and it was this reality that Miss Rosa found Clytie sealing off from her at the foot of the stairs: Sutpen's Haitian son was Charles Bon, and his presence in Yoknapatawpha County meant, as it had in Haiti, that Sutpen's "design" would be "voided and frustrated" (*AA*, 254, 262–63).

As Sutpen struggled to explain his life to General Compson, the Manichaean quality of his innocence emerged. Sutpen had provided for the mother and child, he said; and he had resigned all rights to the property he had acquired with the marriage "in order that I might repair whatever injustice I might be considered to have done." Where had he gone wrong? "You see," he told Quentin's grandfather, "I had a design in mind"; but he could not see how he was responsible for anything else. "Whether it was a good or a bad design is beside the point; the question is, Where did I make the mistake in it[?]" What Sutpen could not see was that his "mistake" was his "design." Spawned in his shock at discovering a world in the Tidewater that seemed the antithesis of his mountain innocence, the "design" allowed him to see the world only in terms of that innocence. Cavaliers like General Compson had inherited a tradition of *noblesse oblige* that allowed them to accept the responsibility for their own and other people's mistakes; but Puritan Sutpen could not see that he owed the woman anything, and Quentin's grandfather could scarcely believe what he was hearing. "What kind of . . . pureblind innocence," he wanted to know, "would have warranted you in the belief that you could have bought immunity from her in no other coin but justice?" (*AA*, 264, 263, 265). Conceived in the Manichaean self-righteousness that allowed him to cast himself only as the exemplar of light, Sutpen's "design" assured its own destruction by the powers of darkness.

Quentin's and Shreve's insights infused Sutpen's story with a sense of reality sorely missing in Miss Rosa's and Mr. Compson's accounts, and, as Sutpen's tale swelled to tragic proportions, the secret that had baffled the hackwriter-narrator of "Evangeline" slid into focus. A father who had rejected his own son because of the shibboleth of miscegenation, who had faced the spectre of incest when that deed came back to haunt him—a father who, finally, had lost both of his sons because he refused to accept one: such were the realities Quentin had been avoiding as he clung blindly to the images of the Old Order that he projected onto Caddy and his grandfather. But Faulkner had not yet finished with Quentin. For as Quentin and Shreve lay pondering the story, Sutpen's tragedy would not die. There were more realities to Clytie's secret than Miss Rosa's devil-planter, than Mr. Compson's redneck *arriviste*, or than Quentin's tragic Manichaean, and the final tale Shreve and Quentin were about to piece together was greater than any of those and transcended them all. For if there had been a dark brother as well as a dark sister in Sutpen's house, Nathaniel Burden's malediction that "the curse of the white race is the black man" was only the tip of the iceberg. Sutpen had been suffering under a curse worthy of the Old Testament God who punished the sins of the fathers to the third and fourth generations; and the South Faulkner was now uncovering was beginning to sound less and less like the South of his youth.

II

Mulatto was a white man's word; it meant little to American blacks, who generally had some white ancestry, except as an example of the familiar talent whites had for mislabeling people. But in the propaganda wars the word had had a long history. Propagandists for abolition developed the tragic mulatto into a familiar stereotype: a slave who inherited strength and penetrating intelligence from a patrician father. Such a slave might enjoy the protection of that father for a while, but in a partitioned society the slave was trapped. Perhaps his father's spirit welled up in him, motivating an escape to the North; perhaps, if escape was impossible, he chose a heroic death to ignominious bondage. In either case, his story was effective as propaganda, particularly with white readers who found it hard to identify themselves with black protagonists. The real paradox of the tragic mulatto, black critic Sterling A. Brown wryly suggested years later, was that it was "a crude kind of racism"; if those "near-white characters" were "the intransigent, the resentful, the mentally-alert" among the fictional slave population of the period, that was "for biological, not social reasons."[4]

Southern writers took a predictably different approach. Many slave-owners assumed mulattos were inferior to the "purely" bred of both races —a "lower order of man," as one critic put it.[5] That notion gave birth to fears that mixture could lower a superior race to the level of an inferior race or, worse, that the weaker characteristics of each might combine to produce a degenerate people. William Gilmore Simms, though far from dogmatic on racial mixture among non-Africans, thought mulattos "a feebler race than the negro, and less fitted for the labors of the field."[6]

Later writers like Thomas Dixon, Jr., foresaw a terrifying future: a yellow nation that would lose the eminence its Anglo-Saxon virtues had brought it. "There is enough negro blood here to make mulatto the whole republic," he warned in *The Leopard's Spots.* "The beginning of Negro equality . . . is the beginning of the end of this nation's life."[7]

Faulkner had been toying with that apocalyptic Puritan vision as early as 1928.[8] Old Doom, patriarch of the Chickasaw Indians, Faulkner reported in "Red Leaves," had a part-black son, Issetibbeha, by a West Indian woman. This unaggressive man had mated with a black slave, and their child, Moketubbe, suggested the kind of fears on which Dixon played. There was something wrong with Moketubbe; it was as though, after two generations of interbreeding, Nature had somehow gone sour, turned on herself. This "copper"-complected man "was maybe an inch better than five feet tall," Faulkner reported, "and he weighed two hundred and fifty pounds." His only ambition seemed to be the wearing of a pair of red slippers his father had brought from Paris; they were hopelessly too small, but when Moketubbe managed to cram his bulging feet into them his face wore an "expression, tragic, passive and profoundly attentive, which dyspeptics wear" (*CS,* 325, 335).

Had miscegenation done that? Miscegenation, those stories showed, was the result, not the cause, of the curse Doom brought onto the land. The curse had originated when Doom set himself up as the owner of land and people; and, significantly, whatever connection Faulkner intended between Doom's moral degeneracy and Moketubbe's genetic degeneracy was never explored. But, as Quentin and Shreve struggled to understand Clytie's secret, Faulkner seemed ready to come to grips with the relationship between moral and genetic degeneracy. The "bitterness" that Quentin had "projected on the South in the form of hatred of it and its people" seemed to have to do with the kind of racial fears Dixon evoked. It had taken four generations for Sutpen's curse to bring about the weakening Doom's blood had undergone in three: the generations of Bon and Clytie, of Bon's son Velery, and of Velery's son Jim Bond.

Clytie's secret had many faces. Charles Bon, handsome and cultured, seemed the genetic and social antithesis of Moketubbe. But, as Quentin and Shreve attempted to trace the man through the legend created by those who had known him, his essence seemed even harder to recapture than Sutpen's. For years, the existence of this dark brother had shaped the Sutpens' lives; but now his only historical remains were a wartime letter addresed to Judith—an ironic note that revealed no more than a weary soldier pining for a deferred wedding. It was as though Bon were "a myth," Mr. Compson told Quentin, "some effluvium of Sutpen blood and character" that the Sutpens had "engendered and created whole themselves" (*AA*, 104). That could explain what fascinated Henry, Judith, and Ellen about Bon: unaware of who he was, they projected their deepest wishes onto him, then fell in love with them, not the man. For Ellen, having that sophisticated New Orleanian for a son-in-law was a crowning social achievement; for provincial Henry, Bon was the worldly wise older brother for whom he had pined; for Judith, he was the brilliant marriage girls dreamed about. Even youthful Rosa Coldfield—younger than her middle-aged sister's children—fantasized a romance with this man she was destined never to meet and lived it vicariously through Judith. That strange name that Sutpen had given Bon seemed to say it all. "Charles Bon, Charles Good," Miss Rosa reflected, and she embarrassed Ellen by speaking aloud what the rest of them wanted to believe: "We deserve him" (*AA*, 148, 76).[9]

For Mr. Compson, who did not yet know about Bon's African blood, these projections identified the tragedy Bon stood for: the rejected brother represented the best in the Sutpen family's aspirations. But Quentin and Shreve were becoming aware of a new dimension to Mr. Compson's Bon. Quentin's father was reacting the way the Sutpens had; his Bon had begun to stand for what Mr. Compson had missed while growing up in the same harsh Yoknapatawpha atmosphere. Compson's Bon was a world-weary Latin "surrounded by a sort of Scythian glitter," older than those country people "not in years but in experience." There was "some tangible effluvium" about him "of . . . satiations plumbed and pleasures exhausted." Amid that world of provincial banality, he comported himself the way Mr. Compson wished he could: with a "fatalistic and impenetrable imperturbability" (*AA*, 93, 95, 94).

That was part of Clytie's secret, too, Quentin was learning: the more one peeled away those masks, the more one seemed to be looking at one's own face. Quentin and Shreve struggled to avoid the snares that had caught the others, but their Bon sounded a lot like Quentin and Shreve, too. Far from their own homes, those young men wanted to know what Bon's childhood had been like; for Quentin, who in *The Sound and the Fury* had blamed his failures on his own mother, that had a special meaning. Bon, they decided, would have known little affection as a child; that deserted Haitian woman would have been charged with hatred, her son no more than an instrument of revenge. The boy would have realized, finally, that his mother was "grooming him for some moment that would come and pass," and then "to her he would be little more than so much . . . dirt." He would have grown up bitter; she would have spoiled him by trying to buy his allegiance with money, and Bon would have hated her for that. Even that morganatic marriage, such a brutal parody of her own legal one, might have grown out of his hatred. "Men seem to have to marry some day," they imagined him taunting her. "And this is one whom I know, who makes me no trouble" (*AA*, 306, 308).

That was a more complex Bon than Mr. Compson's "fatalistic and impenetrable" Latin, and it accounted, as Mr. Compson's Bon did not, for Bon's response when he found his father. If Bon was no more than the instrument of his mother's revenge, all he needed to do was tell Ellen who he was; he had no need to involve Henry and Judith—no need, finally, to die at Henry's hand. But if Bon had rejected his scheming mother, that made all the difference. Bon had probably been dreaming about his lost father all his life, and he might not even have wanted public recognition; this complex man might have wished for something simpler: "it was just the saying of it," and "the physical touch even though in secret . . . of that flesh warmed . . . by the same blood which it had bequeathed him" (*AA*, 319). When Bon was denied that simple thing, Quentin and Shreve decided, his motives had become as complex as his character.

Starved for affection, Bon would have begun to love Judith first as a sister; but, when Henry and Ellen had thrust the girl on him, his libertine past would have surfaced. Shreve insisted to the horrified Quentin that Bon would have found incest fascinating. His octoroon wife excepted, Bon's affairs with women would have been all too easy to slough off; but, "if there were sin too," Shreve imagined him thinking, "maybe you would not be permitted to escape" (*AA*, 324). For Quentin, who in *The Sound and the Fury* had dreamed of such a commitment with Caddy, it was all too convincing; Bon had not planned the threat of incest as a weapon against his father, but he had drifted into it, his motives a strange mixture of love, lust, and the desire for revenge. Thus, when Sutpen tried to stop the wedding by telling Henry—without mentioning Bon's African blood—that Bon was his brother, Bon had taken Henry away to New Orleans; Henry might still be persuaded to accept an incestuous marriage, and Bon now wanted that as much as he wanted recognition. When the war came, Bon was still content to wait: with all three men fighting, a bullet might resolve the issue either way, and by now he did not care which.

The proof of those motives was what happened as the war neared its close. Sutpen, aware that when the Army of Virginia dispersed, Bon would return to carry out his threat, made a last, desperate move: summoning Henry to his camp, he told the boy about Bon's African blood. For Henry, that was traumatic; but for Bon, it offered a potentiality that was fully as terrifying, for Bon would have to come to grips with the truth about his place in that family. If Henry were still ignorant of Bon's African blood, he might yet have allowed an incestuous marriage; aware of it, he never would, and Judith's reaction would be no different. They would see in their dark brother only the social catastrophe every plantation family dreaded.

"He didn't need to tell you I am a nigger to stop me," he said to Henry in camp that night, and he coldly handed his brother a pistol. Henry was shocked. "You are my brother," he protested; but by now Bon at least knew the pathos of his situation. "No I'm not," he replied. "I'm the nigger that's going to sleep with your sister. Unless you stop me, Henry" (*AA*, 356–57). Bon had loved those simple people, Quentin and Shreve decided, but he hated their unyielding rigidity, and he returned to Sutpen's Hundred ready to face his brother's bullet. Fully aware of the ambiguities of his gesture, Bon was a victim of his own tragic complexity. He had no desire to die, as Joe Christmas died, hating his persecutors as he was hated by them; instead he was going to die loving them. But he was not going to allow his family to avoid the consequences of their racism; forcing Henry to kill him meant Bon was placing the responsibility for that racism, finally, where it belonged.

In the threadbare stereotype of the tragic mulatto Faulkner had found the materials for a tragedy that transcended its origins in every way. Bon, rejecting his black identity because he possessed his father's proud spirit and ready to face his fate bravely because of that spirit, sounded superficially as though he could have stepped from the pages of an abolitionist novel; but there was more to Bon than that. Faulkner had modified the abolitionists' stereotype to accentuate its ironies. Unlike men such as Harriet Beecher Stowe's George Harris or Richard Hildreth's Archy Moore, this spoiled young man had never had to face the humiliations of bondage; and it would have been a rare slave who, once provided with the playboy's life Bon enjoyed in New Orleans, might have sacrificed it for the esoteric satisfactions of acknowledgment by a rural patriarch like Sutpen. That not only made Bon more human, it made him an enigma; and Faulkner had seized on another irony that had escaped his abolitionist predecessors, too: Bon's part-black mother. Writers like Stowe and Hildreth, for all their hatred of slavery, had assumed that whatever was cultured in their heroes came from the white, rather than from the African, side of their heritage; Faulkner was more subtle. Redneck Sutpen was not likely to have raised a son of Bon's sophistication, and thus the grace and *savoir faire* that had charmed Henry and Judith came from Bon's mother, with her mixture of black and white aristocrat's blood. But even in the tragedy that the abolitionists had made their own, Faulkner had transcended his sources, for if Bon was tragic, it was not merely, as with the heroes of the abolition novelists, because of the anomalous social role society forced on near-white sons who wanted equality with their white fathers. Here, finally, Faulkner learned to make full use of Gavin Stevens's speculations in *Light in August* that Joe Christmas's "blood would not be quiet, let him save it." If Bon hated his free, near-white mother and loved his octoroon slave wife and her son, if he loved and hated his white father, Stevens's notion of a war in the mulatto's blood might suggest more about Bon than it suggested about Joe. As Quentin turned to the consequences of Sutpen's failures for Bon's dark son and dark sister, this possibility would make all the difference.

III

For writers like Stowe and Hildreth the tragic mulatto showed that the curse of slavery was a moral one; the sins of the fathers were visited on their children, who became tangible symbols of those sins. But such an understanding of the moral implications of miscegenation was not adequate for Quentin; it allowed him to explain "the nigger that's going to sleep with your sister" rationally but left unexplored the apocalyptic redneck fears that people like Vardaman and Dixon played on. For all his Gothic impact the tragic mulatto might serve the same regressive purpose as the Cavalier stereotypes Quentin had projected onto Caddy and his grandfather: to hide, not illuminate, the reality behind his fears about his patrimony. If Quentin was going to come to grips with Clytie's secret, every mask had to be pulled away; he had to grapple with the genetic as well as the moral consequences of Sutpen's curse.

Quentin foresaw that the consequences implicit in Bon's story ought to have been manifested in the other mulattos whom Sutpen had spawned. The most obvious example should have been Bon's son, Velery. Bon's frail slave wife was soon dead, and Clytie brought Velery to Sutpen's Hundred, a gesture that seemed promising: with Sutpen now dead at Wash Jones's hands and Henry still in exile, the family that had rejected its dark brother was making a gesture at expiation by acknowledging his dark son. But Sutpen's curse would not be exorcised so easily. Shattered by what miscegenation had done to their family, Judith and Clytie were destined to influence the boy in all the wrong ways. Raised by these spinsters, Velery learned the meaning of his mixed blood from the two people who had the most reason to fear it. As he grew up, Clytie, terrified that he might sire another mulatto Sutpen, kept him isolated from all human contact outside the family; it was as though she "would have made a monk, a celibate, of him," Mr. Compson said. But what shocked Quentin's father was that it had fallen to one of those women to tell Velery "that he was, must be, a negro." Quentin's father had not found out who had told the boy, but not long after he had learned that one of them had found a mirror hidden under Velery's mattress: "and who to know," he mused, "what hours of amazed and tearless grief he might have spent before it?" (*AA*, 200, 198, 199).

As Velery matured, the results of his training became evident. He became a troublemaker. After he was hauled into court for starting a fight at a black party, he found a promising friend in General Compson, who gave him enough money to leave town. Velery was white enough to pass, the General reminded him, and he could get a fresh start: "What ever you are, once you are among strangers . . . you can be whatever you will." But Velery could not be what he willed. Charles Bon, who passed for white, had died trying to get his white father's acknowledgment; Velery, outraged at what white blood had done to his father, wanted to be black but was too white to pass. A year later Velery was back with "a coal black and ape-like" wife, Mr. Compson said, a woman so feebleminded no one knew whether she was aware her husband was part-black. Like the youthful Joe Christmas, Velery thought black meant bestial; he "had not resented his black blood so much as he had denied the white," Mr. Compson said (*AA*, 204, 205, 207). But Velery's tragedy ran even deeper. Bon harbored no prejudice, and Sutpen saw life only in terms of his "design"; Velery hated both races. Sutpen's curse also appeared to be a social one that allowed Bon and Velery to destroy themselves and their families through compulsive behavior passed on from generation to generation.

Was that Clytie's secret? "Coffee-colored" Clytie herself seemed to be the key to the secret, and she was also its most intriguing riddle. Quentin had encountered Clytie twice, and he had learned more about her from his father and Rosa Coldfield. But the problem was the same one that he had encountered with Sutpen and Bon: everyone seemed to have his own Clytie, and as soon as a person understood one, another emerged. Miss Rosa had projected onto Clytie the full force of her fears about the Cavalier way of life. Clytie, "in the very pigmentation of her flesh," Rosa told Quentin, "represented that debacle [the Civil War] which had brought Judith and me to what we were" (*AA*, 156). But beneath Miss Rosa's demonized projections another Clytie was visible, one who resembled the tragic mulattos of the abolitionist novels. Clytie never talked about her degraded slave mother and never seemed to think of herself as black; she regarded herself as a Sutpen deprived of full status by the accident of a slave mother. Like the mulattos of those old novels, she had inherited her white father's fierce spirit, and it seemed to motivate her brave struggle against the tragedies of birth. But beneath the demonized Clytie and the tragic-mulatto Clytie, another, more human Clytie could be perceived.

The wildness that made Miss Rosa think Clytie bestial seemed more nearly hysteria; it stemmed from a deep sense of outrage at recurring injuries to her Sutpen pride and to those she loved. It showed that Clytie, too, had inherited her father's curse; like him she passed it on to others, transmitting her own tragic ambivalence to Velery. But Clytie had also achieved an ironic stability that no other Sutpen had attained; more enduring than all of the other Sutpens, she spent a long life protecting them. In the novel's closing pages, she emerged as a major tragic figure. The years had taken their toll. The little woman who had stopped Miss Rosa at the stairs years ago was "more than eighty years old" now; but when Rosa brushed past her she was still ready to lay down her life to protect the secret of that dark house. "Make her go away from here," she begged Quentin. "Whatever . . . [Henry] done, me and Judith and him have paid it out." When Rosa came back with an ambulance that Clytie mistook for a police car, Quentin was at last confronted with the scene the hackwriter-narrator of "Evangeline" had witnessed: to seal off its secret, Clytie burned her father's house—herself and Henry in it—to the ground. Quentin remembered that "tragic gnome's face" looking out through the flames, "perhaps," he speculated, "not even now with triumph"; but Clytie might have achieved more than that. Her face was "possibly even serene" now, he believed, there "above the melting clapboards" (*AA*, 351, 370, 376).

For Quentin, Clytie's serene face should have been the final clue to the secret she was guarding. Bon and his pathetic son were driven men; Clytie, caught in the same tragedy, had learned the powers of love and devotion. Quentin should have listened when she told him "me and Judith and him have paid it out." The words revealed the final truth about Sutpen's curse: Gothic enough to satisfy the most diehard abolitionist, the curse resolved itself through Clytie's strength into a daylight world where the acts of responsible people like Clytie had meaning. That was the lost world Quentin had tried to preserve in *The Sound and the Fury* by projecting it onto Caddy and his grandfather, and finding it should have freed him. His failure to find that freedom identified the problem Quentin—and Faulkner—had been struggling with in *Absalom, Absalom!*: the "bitterness" Quentin had "projected on the South in the form of hatred of it and its people." Quentin's projections would eventually destroy him, and they revealed the flaw in his aristocratic commitments. For now that Quentin had accepted one Puritan nightmare about his patrimony—the nightmare of the abolitionists—he was about to have another one forced on him—the nightmare of the Burdens and Dixon.

Sutpen, robust, aggressive, and dynamic, had produced Bon, sensitive, physically hardy, but passive; Bon, allying himself with the frail octoroon, had produced the equally frail Velery. Clytie's black Haitian mother had been as robust as Sutpen, but Clytie, who had outlived all the Sutpens of her generation save Henry, was small in stature, and there were other hints of genetic weakness. As Clytie aged, something strange happened. When Quentin had seen her first, he told Shreve, Clytie was over seventy; but she still "had no white hair" and "her flesh had not sagged." Did that mean something had gone wrong genetically? To Quentin, Clytie looked like "she had grown old up to a certain point just like normal people do, then had stopped"—and "instead of turning gray and soft she had begun to shrink, so that the skin of her face and hands broke into a million tiny cross-hair wrinkles and her body just grew smaller and smaller." To Quentin, long before Clytie immolated herself in Sutpen's rotting mansion, she had begun to look "like something being shrunk in a furnace, like the Bornese do their captured heads" (*AA*, 215).

Had miscegenation done that? And if intermixture could do that in one generation, what could it do in four? Quentin himself had seen the product of Velery and his black wife, the creature neighboring blacks called Jim Bond, who howled and ran off into the darkness as people approached: the creature had stared at Quentin, "his arms dangling," and "no surprise, no nothing in the saddle-colored and slack-mouthed idiot face" (*AA*, 370). That regression from Sutpen to Jim Bond was a vision that would produce goose-pimples on a Dixon, and Faulkner allowed its impact to register on Quentin through his outlander, Shreve McCannon. There were Puritans, it appeared, in more places than the Burdens' New England. "The South," mused Shreve, "Jesus," and his horrified reaction made him sound like a Canadian Dixon. "So it took Charles Bon and his mother," he told the shaken Quentin, "to get rid of" Sutpen, and it took "Charles Bon and the octoroon to get rid of Judith, and Charles Bon and Clytie to get rid of Henry." Thus, Shreve ironically concluded, "It takes two niggers to get rid of one Sutpen," and he thought balancing Sutpens and "niggers" like that was "fine"—after all, he said, "it clears the whole ledger, . . . except for one thing." That "thing" was howling Jim Bond. "You've got one nigger left," he told Quentin. "One nigger Sutpen left" (*AA*, 377–78).

To Shreve, this last entry in Sutpen's ledger cancelled the rest; it destroyed any hope that Jim Bond might be no more than a punishment aimed at Sutpen's white family. For if the "niggers" had gotten rid of all the Sutpens, it was also true that the Sutpens had gotten rid of all the "niggers," and all that was left was a "nigger Sutpen": an idiot mulatto, the product of four generations of interbreeding, and, in Shreve's eyes, a reversion to a near-bestial state. "Of course you can't catch him and you don't even always see him and you never will be able to use him" he told the horrified Quentin; and he reminded him with a sneer, "You still hear him at night sometimes" (*AA*, 378).[10]

"You can't understand" the South, Quentin had cautioned his friend. "You would have to be born there." That seemed all too true; but, whatever Shreve's diatribe proved about Shreve, it showed how little Quentin understood those things he himself had projected on the South. Shreve, born farther from the South than the Burdens, had caught a vision that hit Quentin too close to home; recalling the genocidal rantings of southerners like Doc Hines or Dixon's malediction that "the beginning of Negro equality . . . is the beginning of the end of this nation's life," its harrowing conclusion left Faulkner's "protagonist" shaken. "I think that in time," Shreve mused, "the Jim Bonds are going to conquer the western hemisphere." It was not going to "be in our time," he admitted, and like Calvin Burden he thought they would "bleach out." But Shreve was sure that "it will still be Jim Bond," and that implied a consequence that Quentin was not prepared to deal with: "in a few thousand years, I who regard you will also have sprung from the loins of African kings" (*AA*, 378).

Whatever the truths of Clytie's secret, Faulkner had already made it clear that Shreve's genetic apocalypse was not one of them. Jim Bond was no inevitable product of genetic degeneration; he was no more, as *Absalom, Absalom!* showed, than the son of a moronic woman, and a similarly logical explanation could be made for every other development in which Shreve found an omen. Velery's frailty was an obvious result of his mother's frailty, Clytie's strange appearance a normal function of aging; and, significantly, in Clytie, Faulkner had allowed Quentin to observe how the curse Sutpen had invoked could teach one individual the traditional values of love and commitment. But Quentin's problem was not the truth about Sutpen's curse; it was what he projected onto that curse.

Assailed first by his own doubts about his patrimony, then by Puritan Miss Rosa and Puritan Shreve, Quentin now experienced the full impact of the apocalyptic fears to which Vardaman and Dixon gave voice, and he found himself suddenly vulnerable to every doubt he had ever held about his heritage. If these fears were capable of materializing in the Jim Bonds, the nineteenth-century nightmare the Sutpens had lived seemed all too close to the twentieth-century nightmare Quentin was living. Bon's passion for Judith was a brutal parody of Quentin's for Caddy; Clytie's blood loyalty to her white family could only suggest a new and shocking dimension to Dilsey's loyalty to the Compsons; and if miscegenation could produce a Jim Bond—well, there was an idiot in Quentin's family, too. Those shocks left Quentin ready for a fatal reappraisal of his family's past, based on the Puritan, rather than the Cavalier, tradition. If the Sutpens had vanished from Yoknapatawpha, it was clear that the Compsons were on their way out also, and the parallels were all too similar. Had Quentin's grandfather, General Compson, invoked the same kind of curse Sutpen had invoked? Had he been any less a pretender to aristocratic values than Sutpen had been? Had anyone? Those were the sources of the "bitterness" Faulkner had been talking about in 1934 that Quentin had "projected on the South in the form of hatred of it and its people." Shreve did not know what he was asking as he pressed Quentin, "Why do you hate the South?" Trapped in that Manichaean vision, Quentin could only lie panting in his bed in "the iron New England dark," as he struggled to deny the darkness that seemed to be falling on him and his patrimony. "I dont," he told Shreve. "I dont! I dont hate it! I dont hate it!" (*AA*, 378).

IV

As Quentin lay dazed under the impact of this Puritan vision, the South Faulkner had been searching for appeared ready to open up before him. With that vision, Faulkner seemed to have thrust past the two most formidable barriers to the "recapture" of the world he had "been preparing to lose and regret" in *Sartoris*: through Clytie, he had achieved an open-eyed look at the realities of the curse abolitionist Puritans claimed planters brought to their heritage; through Quentin, he had learned to separate himself from the apocalyptic fears that southern Puritans projected onto aristocrats. Those were major breakthroughs for Faulkner, and the measure of their importance was that *Absalom, Absalom!* read like a retrospective commentary on every earlier phase of Faulkner's search for a South. Thomas Sutpen, with his one-generation rise to fortune, suggested the facts rather than the myth about Colonel W. C. Falkner—as John Sartoris never had. The image of aristocracy with which Sutpen's children identified themselves was their family's own creation, fabricated in a single generation; they understood genuine aristocracy no better than other *arrivistes*. This fact implied a great deal about the Cavalier image Faulkner had projected onto John Sartoris; it also revealed much about Quentin's inability to recover those values in *The Sound and the Fury*; and it opened up for Faulkner, at last, the whole question of the realities of slavery. In *Light in August* and the Sartoris stories Faulkner had manipulated his plots to avoid confronting the guilt of the planter class, allowing the blame to fall on Puritan rednecks and Puritan abolitionists; but Sutpen presided over a scene Gothic enough for a Simon Legree. Those Haitian slaves, Mr. Compson reported, had been forced to live naked at Sutpen's Hundred in the early days; the women were used like breeding mares; like Legree, Sutpen had his slaves fighting for sport; like Legree, he bedded with the women.[11]

These were considerable admissions for Colonel W. C. Falkner's great-grandson, and they had an immediate effect on his new novel. The measure of this effect was the distance between Bon and Ringo, between Clytie and Dilsey. The tragedy of Henry Sutpen's dark brother was a telling commentary on the inhibitions with which Faulkner had approached Bayard Sartoris's dark brother; if the darker Sutpen excelled his white counterpart in every way, that was no longer, as it had been for Ringo, a sentimental compliment—it was heartbreaking. And what was true of Bon and Ringo was true of Clytie and Dilsey. *Absalom, Absalom!* provided a reason for Clytie's enduring loyalty; there was no reason for Faulkner to smother her in the fog of sentiment with which he had surrounded Dilsey. Finally, all these achievements pointed to what Faulkner had achieved in Quentin. Quentin had never been fully real to Faulkner, and it had taken Faulkner eight years to exorcise himself of his "protagonist." Now, blinded by the "bitterness" he had "projected on the South and its people," Quentin was gone, victim of a more believable suicide; but an open-eyed Faulkner had found the courage to explore Clytie's secret, and, because he had, his nihilistic "protagonist" would never trouble him again.[12]

That was why, for the moment, *Absalom, Absalom!* seemed to open up unlimited vistas. If Faulkner had at last learned to separate the failures of the world of his youth from its redeeming virtues, the prospect of a visionary South was opened up, a South where the values John Sartoris stood for could be recycled to free southerners from the curse that had destroyed Sutpen. But Faulkner had reason to reflect now on how such goals had eluded him in the past, and there remained hints of that past in *Absalom, Absalom!* Fascinated by his mulattos, Faulkner had scarcely bothered to include *any* other kind of black in his novel; no character who identified himself as basically African caught his attention. Faulkner had once again avoided the meaning of growing up black in a white-dominated society, and this had a direct bearing on the authenticity of his part-black figures. Many nineteenth-century mulattos, no doubt, identified themselves with white fathers; but Faulkner's mulattos did that with such consistency, and with such disregard for their black ancestors, that the pathos of their tragedies was ignored. Faulkner had inherited the problem, perhaps, from the white abolitionists who created the tragic-mulatto stereotype. Still, he left himself open, as they had, to Brown's charge that they were practicing "a crude kind of racism." Like the abolitionists' mulattos, Faulkner's figures were in no way representative of the majority of slaves. Field hands forced to fight each other like animals, women forced to breed like mares—those were the tragedies behind the bland mask of paternalist propaganda; but Faulkner's preoccupation with Bon, Clytie, and Velery obscured the problems of their fellow blacks more than,it explained them.

Faulkner's debt to the abolitionist tradition may have been responsible for another area where his vision faltered. Stowe, struggling to keep southern readers from escaping responsibility for their beliefs, made her devil-planter Legree a Yankee to prove it was the system, not southerners, who made slavery vicious. Faulkner could have been thinking of Legree when he went outside the slaveholding South to create Sutpen, a mountain man who had little more understanding of paternalism than Stowe's villain. The results had been suggestive for his novel: making his devil-planter a species of redneck had helped Faulkner to separate the vices of the system from the character of southerners who were locked into it. But for Faulkner, the choice had also had an opposite effect: it allowed him to attack those vices without attacking the basic assumptions of the aristocrats who had created the system. *Uncle Tom's Cabin* argued that any system based on slavery was evil: *Absalom, Absalom!* argued that this was true only when such a system was not run by aristocrats.

Such manipulations were familiar from Faulkner's earlier works; in *Light in August* and in the Sartoris tales they had become virtually mannerisms. The irony of Faulkner's new novel was that it did not allow him to push much closer to the realities of the world of his youth than these earlier stories had. Making his plantation patriarch a redneck *arriviste* should have brought Faulkner closer to those realities; but he had neatly sidestepped them by projecting the ideals of that world onto General Compson and the Puritan fears about those ideals onto Sutpen. Those manipulations boded ill for the future of his search for a South; there, planted firmly at the roots of Faulkner's triumph, were the seeds of his tragedy. Quentin, muttering compulsively that he did not hate the South, had at least been ready to face the worst about Clytie's secret. But Faulkner had manipulated his plot so that he did not have to, and the effect on his career was going to be more devastating than if he had never tried to deal with that secret at all.

9

Letting Go

I

When strangers asked Faulkner in the 1940s and 1950s about his life-style, they did not always hear what they expected. "I think of myself as a farmer, not a writer," he would tell people. He seemed proud of the farm he had purchased two years after *Absalom, Absalom!* appeared. "My life is farmland and horses and the raising of grain and feed."[1] As critic Malcolm Cowley pieced together what Faulkner told him about that farm, it sounded more like a small plantation. "Apparently his farm . . . is a large one," Cowley decided, "with part of it under cultivation by three Negro families." Faulkner's attitude toward the farm interested Cowley; his friend seemed to have given up trying to make money out of his tenants, and Cowley thought he knew why. Faulkner, "speaking very softly," had confided that "the Negroes don't always get a square deal in Mississippi."[2] Those who knew Faulkner better wondered why he kept the farm. John Faulkner thought it was because his brother had a "tremendous love for his native soil, and it pleased him to be identified with it." Jack Falkner was more direct; the farm was "unself-supporting," he said, and Bill "wrote to support it." Both might have agreed with John Cullen that the real reason was closer to home. "He keeps the farm because he . . . wishes to think of himself as a farmer," Cullen wrote. "For him, farming is actually a hobby."[3]

That farm was scarcely as Gothic as Rowan Oak, around which Faulkner had erected his edifice of myth about Sutpen, and it was insignificant in size compared to Sutpen's Hundred or to the sprawling plantation he would soon write about; but it provided Faulkner with a vehicle for the resurrection of still another memory from the world of his youth. The man he resembled most was not the Old Colonel but J. W. T. Falkner, with his own small farm north of Oxford and his bank and political commitments that made him independent of it; and, as Faulkner struggled to protect those families who "didn't always get a square deal in Mississippi," he could speculate, at least, about how both his grandfather and great-grandfather had dealt with such problems.[4]

Faulkner's retinue at Rowan Oak, meanwhile, was showing signs of attenuation. Redoubtable Ned Barnett had found the farm more to his liking; established as one of its tenants, he occupied himself with selling Faulkner his cotton before harvest so that the man he still called "Master" would have to pay for getting it picked.[5] And Mammy Callie was gone, victim, at age one hundred, of a stroke. When Mammy's time came—it was January 21, 1940—the Faulkners laid her out as she had asked them, in her apron and cap, and Faulkner, fulfilling a promise, preached the funeral sermon. The inscription he placed on her gravestone indicated his devotion[6]:

Callie Barr
Clark
1840–1940
MAMMY
Her white children
bless her

Two years later, when *Go Down, Moses* appeared, the legend on its flyleaf resembled the one on Mammy's grave:

To Mammy

CAROLINE BARR

Mississippi
[1840–1940]

Who was born in slavery and who
gave to my family a fidelity without
stint or calculation of recompense
and to my childhood an immeasurable devotion and love

Go Down, Moses was going to be another monument to Mammy Callie, and it would be a monument to other familiar memories, too. Ned Barnett would be there, and there would be a farm—a farm far larger than Faulkner's or his grandfather's, but one whose dependents bore some resemblance to the Faulkners' own.

Writing *Absalom, Absalom!* had been a considerable concession for the Old Colonel's great-grandson. Many planters, the novel implied, had been more like Sutpen than like John Sartoris. That meant that abolitionists had been right in much of what they said about his patrimony, and if that were true it placed blacks like Mammy Callie in a different perspective; it meant that, as Faulkner preached Mammy's funeral in

1940, he was already aware of a new direction his search for a South was taking.

When Faulkner had brought out his Sartoris tales in 1938 as *The Unvanquished,* he had added a new section to "Retreat" and written a final story, "An Odor of Verbena," which tied the series together. As Colonel John built his railroad in the latter story, Drusilla Hawk, now Colonel John's wife, tried to tell Bayard what paternalism ought to mean; the Colonel, she said, had a dream, and his son ought to appreciate it. Dreams made twenty-year-old Bayard think of Sutpen, "a cold ruthless man" and "underbred": "nobody could have more of a dream than Colonel Sutpen." Drusilla wanted Bayard to understand the difference between Sutpen's dream and Colonel John's; it was *noblesse oblige.* "But his dream is just Sutpen," she protested, "John's is not," and, as she struggled to explain what Colonel John's dream was, it sounded like Granny Millard's. What John Sartoris was working for was "not just his kind . . . , but all the people, black and white, the women and children back in the hills who don't even own shoes"; Colonel John was "thinking of this whole country which he is trying to raise by its bootstraps" (*U,* 168–69).

That was a different Colonel John from the Cavalier patriarch of *Sartoris.* The "stubborn dream" that was still haunting his descendants in that novel had little to do with any benevolent paternalism; it sounded more like Sutpen's. But in 1938, with *Absalom, Absalom!* behind him, Faulkner was ready to add a new dimension to Colonel John: he had elevated his Cavalier to the status of a moral symbol. That had the effect of balancing the critique of paternalism implied in the earlier novel and giving it perspective, as did the sections Faulkner added to "Retreat" about a pair of landowners who sounded even more countrified than Sutpen. Amodeus (Uncle Buck) McCaslin and his twin brother Theophilus (Uncle Buddy), Colonel John had told Bayard, "were ahead of their time" because of their "ideas . . . about land," which were radically different from those of their neighbors. "They believed that land did not belong to people but that people belonged to land and that the earth would permit them to live on . . . it and use it only so long as they behaved and that if they did not behave right, it would shake them off just like a dog getting rid of fleas" (*U,* 45).

These twins seemed aware of the curse Sutpen had invoked, and they were struggling to avoid it. For example, they had set up, even before manumission, "some kind of a system of bookkeeping . . . by which all their niggers were to be freed." This was not to be the too-easy freedom that frustrated the River Jordan marchers. The McCaslins' blacks were "not given freedom" outright; they got it by "earning it, buying it not in money from Uncle Buck and Buddy, but in work from the plantation."

That was a different order of relationship to the land from Sutpen's; blacks owed their work not to the Man but to the land itself. But more significantly, Buck and Buddy had put together an interim plan to deal with the situation that existed before the blacks earned their freedom. From their father the twins had inherited "a big colonial house," but they had never lived there since their father's death. "They lived in a two-room log house with about a dozen dogs," Bayard learned, and "they kept their niggers in the manor house." For wartime Mississippi, that sounded like madness, but there was more. Every night the twins and their blacks went through a strange ritual; ceremoniously, Buck and Buddy locked the blacks in the mansion, but "probably," Bayard thought, "they would still be locking the front door long after the last nigger had escaped out the back." If people belonged to the land, not the land to people, Buck and Buddy knew, slavery was wrong, and that strange ritual was the way they had adapted to it. Blacks, symbolically enough, were established in the manor house; that was one way of setting things straight. But they were not really straight until the blacks were free of the house, too, and, under the laws of slaveholding Mississippi, this ritual was what the twins had invented. "It was like a game with rules," Bayard thought, and it sounded like a compromise with which people could live. The twins were no Manichaean abolitionists like the Burdens; they had come to terms with the darker and lighter sides of their natures, and they had found a kind of peace. Every night they sat in their log house with their dogs—playing poker, people said, "betting niggers and wagon-loads of cotton with one another on the turn of a single card" (*U*, 45, 44).

These new additions to Faulkner's Sartoris tales showed that he was still manipulating his plots, as he had since *Light in August,* to make the Civil War look like a Yankee mistake. Faulkner wanted his readers to believe that, in this countrified setting that made Sutpen's Hundred sound like a Delta estate, the seeds of southern reform were being sown even before the Civil War; and, if aristocratic Colonel John was able to understand what country reformers like the McCaslins were doing, then his entire class was capable of self-criticism. The implications of both facts for Faulkner's search for a South were sweeping. If the South had been moving toward the new awareness represented by the McCaslins, the Civil War and Reconstruction had been catastrophic blunders; rather than speeding manumission, they had been obstacles to a process that would have come about sooner if outsiders had not interfered. But if Faulkner seemed to be moving away from the insights the abolitionist myth had allowed him in *Absalom, Absalom!,* this new material showed he was still preoccupied with the curse Sutpen had invoked.

What frightened Bayard about his father's dream was the amorality with which Colonel John sometimes pursued it; he had developed a reputation for involving himself in duels with people who got in his way. Bayard found that hard to live with, even when his father's victims were carpetbaggers like the Burdens. "They were men," he told Drusilla. "Human beings" (*U*, 169).[7] The attenuation of Colonel John's character that Bayard had identified was of a different order from Sutpen's; Colonel John was not driving himself to finish his railroad out of a mere desire for self-aggrandizement, but out of a paternalistic concern for the people of Yoknapatawpha County. Faulkner was still unwilling to deal with any guilt that the Sartorises had acquired in their relationship with Ringo and the rest of the black Strothers. Still, it was quite a concession for Faulkner; for the first time he dealt directly with the possibility that a blooded aristocrat might suffer from anything resembling the curse Sutpen had invoked.

But it was not so much Colonel John's guilt as Bayard's reaction to that guilt that showed where Faulkner's search for a South was taking him. Aware that the code his father had lived by required his son to avenge his death, Bayard went to meet Redlaw, his killer; but he shocked everyone by refusing to carry a weapon. For a traditional aristocrat like Drusilla, that was unthinkable; it was an affront to all Colonel John had stood for, and, even after Redlaw had fled, unwilling to live with the deaths of two men on his conscience, she could not accept it. But others, such as Miss Jenny Du Pre, understood better; even George Wyatt, Colonel John's wartime aide, could see the need for change. "Maybe you're right," he told Bayard, "maybe there has been enough killing in your family" (*U*, 189). With Bayard's gesture the new direction of Faulkner's search for a South slid into focus. From one standpoint, that gesture served the same purpose that the McCaslin twins' theories about land had served; it suggested that the aristocratic class had been capable of dealing with the curse that harrowed it and hence that the Civil War had been a catastrophic blunder. But like the twins' decision to move out of their mansion, it was also an acknowledgment of the truth of the abolitionists' accusations and an attempt to respond to them. The twins, through their scheme that allowed slaves to earn their freedom by working for it, had found a way to resolve their guilt about their heritage; Bayard, by facing his father's killer unarmed, had solved the problem another way. Together, those two events represented a landmark in Faulkner's search for a South.

The Hamlet (1940) gave no hint of that. At last Faulkner had brought together a volume of the Father Abraham material, but somehow Flem Snopes—that slimy representative of the rise of the rednecks—seemed less menacing now, caught in the complex tapestry of one of Faulkner's most stylized works. The episodic novel he published two years later, dedicated to Mammy Caroline Barr, was the logical successor to *Absalom, Absalom!* and *The Unvanquished.* In *Go Down, Moses,* Faulkner embarked on a new phase of his search for a South: he was ready, at last, for his effort to find that visionary South where the paternalistic values Colonel John had stood for might be recycled to redeem southerners from the curse people like Sutpen had invoked. *Go Down, Moses* was going to show how a Cavalier family that practiced the kind of paternalistic responsibility Faulkner associated with John Sartoris might also suffer under a curse like Sutpen's; and it was going to unveil a new hero who, born into such a family, might learn how to free it from that curse.

II

Faulkner needed a new "protagonist." Suicidal Quentin was behind him now,[8] and Bayard, who had faced the guilt of John Sartoris's killings, was locked into an unexamined relationship with Ringo and the black Strothers. What Faulkner needed now was a protagonist who could grapple with every aspect of the curse he had explored in *Absalom, Absalom!,* and so he turned to the family of country slaveholders whom John Sartoris thought "were ahead of their time." In Buck McCaslin's son Isaac, Faulkner was going to fuse Bayard's paternalist outlook with Quentin's tragic revulsion at the curse of inherited guilt; fully steeped in the Cavalier tradition, Isaac would yet be capable of identifying the curse Sutpen had invoked and of translating that knowledge into a personal commitment to the land and its people. Old Buck and Buddy had understood that "land did not belong to people but that people belonged to land," and they had struggled to do something about that; Isaac was going to push those ideas through to their conclusion. Faulkner was not yet ready for a hero who could transcend this curse; Isaac would be more tragic than Quentin, for he would be unable to shape his own life to his vision. But what was important to Faulkner was the effort that Isaac would make. To free his descendants from that curse, Isaac, on his twenty-first birthday in 1888, was going to repudiate the plantation his father had bequeathed him.

This was a remarkable development. The writer who had virtually idealized John Sartoris as the epitome of *noblesse oblige* now suggested that no aristocrat could be free from the curse without giving up everything. As Faulkner filled out the story of Isaac's grandfather, old Carothers McCaslin, the lengths to which he was now ready to go seemed even more revealing. The further Faulkner explored the problem of inherited guilt, the less his plantation patriarchs resembled the Falkner family's official myth. The rise of redneck Thomas Sutpen, who had fought his way up from poverty, resembled the realities of the family myth more closely than the story of aristocratic John Sartoris; but Sutpen, who like Colonel Falkner was bent on absorbing the aura of Cavalier respectability, had held himself ruthlessly to his "design." Old Carothers was the logical resolution of the process begun with Sutpen and resembled Colonel Falkner only in his drive to power. Faulkner's new patriarch was merely gross and suggested that the worst of the abolitionists' fears about planters might be true. A ruthless man who had purchased a vast segment of land from old Ikkemotubbe (Doom) in the early days of the century, he had set himself up as the owner of a sprawling, slave-run plantation; and in that country grandeur he had lost himself in miscegenation and incest as none of the Sutpens had. Purchasing a slave mistress, he had set her up as the wife of an old family slave and fathered a daughter, Tomey, by her; years later, he had fathered a son, Turl, by that daughter. Those were crimes flagrant enough to validate the worst suspicions of the abolitionists: old Carothers, in his rage for self-aggrandizement, had violated both the land and its people.

Carothers McCaslin was a significant development in Faulkner's search for a South: his presence in *Go Down, Moses* suggested Faulkner was ready to face the worst implications of slavery. But the very quality of the McCaslins' life-style presented a problem. Not merely old Carothers but also his forward-looking twin sons, Buck and Buddy, were too much like rednecks; they were too ignorant to have transmitted to their descendants any notion of the old South's official self-image. If Isaac McCaslin's repudiation of his heritage were going to have meaning for the paternalist tradition, there had to be some kind of transfusion of aristocratic sensibility into the McCaslin blood. That, perhaps, was why Faulkner had begun "Was," the first episode of his novel, with a youthful narrator named Bayard;[9] a marriage of McCaslins and Sartorises would have supplied the feeling for tradition that the McCaslins lacked. But Faulkner had already explored the story of Bayard's descendants in *Sartoris,* and before he was through with "Was," Bayard had become McCaslin (Cass) Edmonds, son of a daughter of old Carothers. The Edmondses, it appeared, were aristocratic stock, and it was Cass, a more cultured and articulate man than old Buck, who had run the plantation after the twins had passed on, raising Isaac, the child of Buck's old age, like a son. As Cass tried to explain old Carothers to Isaac in "The Bear," the book's fifth episode, he sounded like a grouchy Bayard explaining a seedy John Sartoris.

For Cass Edmonds, old Carothers had been an aggressive man of vision, ready to accept the responsibilities of power. When Carothers had traded that plantation away from Ikkemotubbe, he reminded Isaac, it had been no more than "a wilderness of wild beasts and wilder men"; what mattered was that Carothers "saw the opportunity and took it," that he "got the land no matter how." His ambitions, Cass was ready to admit, were personal; he had "translated it into something . . . to perpetuate his name and accomplishments." But old Carothers had had his own family in mind, too; he had wanted "something to bequeath . . . for his descendants' ease and security and pride." Edmonds saw no reason those descendants should not accept such a legacy, and he was ready to justify it with a theory of black behavior. Blacks, he believed, were irresponsible, childlike creatures, ravaged by congenital vices: "Promiscuity. Violence. Instability and lack of control. Inability to distinguish between mine and thine." They had to be protected by responsible whites, and so the heritage old Carothers had left was not merely a privilege but also a responsibility. If Edmonds seemed to have done a lot of theorizing about that, he had reason: with the deaths of the twins he shouldered that responsibility. If old Carothers had been a kind of founding father, Edmonds was a kind of Redeemer. He had brought the plantation "through and out of the debacle" of Reconstruction, "where hardly one in ten survived"; and under his supervision it was going to "continue . . . solvent and efficient and intact" (*GDM*, 256, 294, 298).[10]

Isaac had grown up as steeped in the Old Order's rationale as a J. W. T. Falkner. Because of Cass's Reconstruction role, Isaac felt that his cousin was rightly "the head of my family," and Isaac's feelings about the war and Reconstruction ran deep. Southerners had not fought "because they were opposed to freedom," he insisted. They were no different from anybody; they had fought "for the old reasons for which man . . . has always fought and died in wars: to preserve a status quo or to establish a better future one . . . for his children." The proof of that view was the odds they had taken on: who would have tried it "except men who could believe that all necessary to conduct a successful war was . . . just love of land and courage"? Isaac had no sympathy with the northern role in Reconstruction; that had been a "dark corrupt and bloody time." Carpetbaggers were a "race even more alien to the people whom they resembled in pigment . . . than to the people whom they did not"; what singled them out was not a desire for black betterment, but "a single fierce will for rapine and pillage"; their descendants would lead the Ku Klux Klan of a later day (*GDM*, 288, 290, 289).[11] Isaac was a committed southern aristocrat, as loyal to his ethnocentricity as Bayard Sartoris.

What could bring a man like that to recognize the curse he was struggling under? Not, certainly, the Manichaean abolition fever of carpetbaggers like the Burdens. But there was another kind of spirituality in

Yoknapatawpha County, and Faulkner had been aware of it since *Soldiers' Pay.* So far, that spirituality had existed alongside the Manichaean Puritanism of Faulkner's poor whites and abolitionists and the bland high-church attitudes of his aristocrats, the traditions continuing parallel and insular; but Faulkner was ready to experiment with mixing those traditions. The blacks and Indians of *Go Down, Moses* were not altogether like those mystical primitives of *Soldiers' Pay, The Sound and the Fury,* and the Indian stories; Faulkner seemed less interested in them as exotics now and more interested in defining their spirituality. But perhaps because he would get at that spirituality through Indians rather than blacks, this mysticism was at last going to be communicated to whites.

For Isaac's spiritual mentor Faulkner brought back Sam Fathers from "A Justice" (1931), a part-black son of old Doom whom that Chickasaw patriarch had sold to Carothers McCaslin along with his part-white mother.[12] Sam's Indian and African heritages provided him with an awareness of which whites were incapable; at Sam's birth, Edmonds told Isaac, "all his blood . . . except the little white part, knew things that have been tamed out of our blood." When Sam talked to him about the wilderness, Isaac felt that "he was the guest here and Sam Fathers' voice the mouthpiece of the host." That voice spoke for a world in which inherited attitudes provided a man with all he needed to know. Sam's very life, Isaac discovered, was a form of worship. Old Ben, the aging bear they hunted every fall, was the "epitome and apotheosis of the old wild life"; the annual hunts Sam led in which no one expected to kill Old Ben had become a "yearly pageant-rite of the old bear's furious immortality," a ritual of communion with the spirit of the wilderness participated in by men and animals alike. When twelve-year-old Isaac killed his first buck, Sam painted his face with its blood, and Isaac came to realize that the old man had "marked him forever one with the wilderness." It was Sam, he came to understand, not Cass Edmonds, who "had been his spirit's father if any had" (*GDM,* 166, 167, 171, 193, 194, 178, 351, 326).

Sam's African heritage forced Isaac into a soul-searching reappraisal of his inherited ideas about blacks, and the result was a radical theory of black character that changed his life. Isaac was rhapsodic about blacks: "They are better than we are" and "stronger than we are," he decided. They possessed formidable virtues: "endurance" and "pity and tolerance and forbearance and fidelity and love of children." They were vessels of a singular spirituality inherited from "the old free fathers" of Africa. And they had acquired other virtues from the experience of bondage: they "had learned humility through suffering and learned pride through the endurance which survived the suffering." From his experiences with Sam and others, Isaac had conceived a virtual beatitude of black virtues that was as glowing an affirmation of black American character as any in American literature; and it was this romantic interpretation of Sam's African heritage that allowed Isaac to identify the curse old Carothers had brought onto the land. "Sam Fathers," he told Cass, "set me free" (GDM, 294, 295, 300).

White aristocrats, locked into their inherited outlook, had trouble understanding the curse; when they looked for an alternative to paternalism, all they could see was the Manichaean Puritanism that had produced the abolitionist crusade. But Isaac had been able to burrow under that. Committed more to his primitive than to his European heritage, Isaac could see aspects of his white culture that whites could not see. Dimly aware of a blood relationship between his white family and some of its former slaves, Isaac dug through the plantation birth records in the commissary for hints of his grandfather's sexual misconduct; shocked, he struggled toward an agonizing re-evaluation of his family's responsibility in the dispossession of the Indians and the enslavement of the blacks. Faulkner here returned to a device that had served him well in *Absalom, Absalom!*: a debate between his protagonist and a sympathetic opponent over the nature of southern reality. As Isaac sat with Cass in the commissary, "juxtaposed . . . against the tamed land which was to have been his heritage," he struggled to make his cousin understand why he could not accept that heritage.

Isaac had built up an elaborate personal mythology, a strange mixture of the transcendental, the pagan, and the biblical, which he thought identified his historical role. That role was a natural extension of his father's and uncle's belief "that land did not belong to people but that people belonged to land." The curse Faulkner's Puritans had struggled with in *Light in August* and *Absalom, Absalom!* found its most elaborate theorist in this descendant of Cavaliers. Isaac had come to believe that "this whole land, the whole South, is cursed, and all of us who derive from it"; like the Burdens, he was sure that "black and white both, lie under the curse" and that it was inherited to the third and fourth generations. But Isaac's God was a different God from the wrathful God of the Burdens; he was a loving, patient God who used his curse to teach, rather than to avenge himself. He came through, in fact, like a sort of kindly landowner running some cosmic plantation: "He created man to be His overseer on the earth," Isaac thought. But it was God, not man, who owned the earth. God had never meant that an individual man might "hold for himself and his descendants inviolable title forever"; he had intended for all men "to hold the earth mutual and intact in the communal anonymity of brotherhood." In Europe men had violated God's trust, had degenerated in "the old world's worthless twilight" until they "snarled" like animals "over the old world's gnawed bones, blasphemous in His name." In his patience God had revealed America to them; he wanted to give them a second chance in "a new world where a nation . . . could be founded in humility and pity and sufferance and pride one to another" (GDM, 278, 257, 258).

How could a loving God be the source of a curse? Isaac's idea of the origin of the curse was as far a cry from the rantings of the Burdens as his God was from theirs. People like old Carothers had brought the curse on themselves by violating God's "communal anonymity" for their own aggrandizement; but God's vision, Isaac thought, was a beatific one, a divine theory of compensation in which evil worked for good. Concluding that men could "learn nothing save through suffering," God had supplied them with enough suffering to understand the meaning of the land he had given them. For Isaac, God's method was thus an evolutionary one, and he thought he could see that evolution working even in his own family. "Maybe," he told Cass, "He knew that Grandfather himself would not serve His purpose . . . , but that Grandfather would have . . . the right descendants." Isaac understood his father's and uncle's moral vision that had led John Sartoris to think they "were ahead of their time." But old Buck and Buddy were only one stage in that evolution. In realizing that "people belonged to land," they had made their contribution, but the twins had never come to grips with the worst of their father's misdeeds. Their black brother—the son of their black sister Tomasina, or "Tomey"—had been acknowledged only through a shabby legacy of some money and land; he had never been allowed to use the name McCaslin. He was known only as "Tomey's Turl," and in "Was," Buck called him "my nigger" and chased him with dogs when he tried to escape (*GDM*, 286, 259, 11). Despite the twins' efforts to heed God's will, their father's guilty heritage had been passed on to the next generation, and the heir was Isaac.

Isaac believed that he was too close to that guilt. At fourteen, he recalled later, he had believed he could "cure the wrong and eradicate the shame"; but at twenty-one, he "knew that he could do neither." Sam Fathers might have freed Isaac from his commitment to that heritage, but in a profound sense he was never going to be free. He "could repudiate the wrong and shame," he knew, "at least in principle, and at least the land itself in fact"; but the best Isaac could hope for himself was that somehow he could escape God's curse. If he did, his descendants might have a chance for freedom; that was what Isaac meant when he spoke of how it had taken God "three generations . . . to set at least some of His lowly people free": what Isaac was doing, he was doing for his son (*GDM*, 351, 259).

Perhaps most remarkable about Isaac's notions was his view of who those "lowly people" were going to be. They were not the people old Carothers had taken the land from—after Sam's friend Joe Baker died, Sam was the only Indian still there; and they were not the people old Carothers had brought there to work the land either. Some of these had vanished like the Indians; of old Turl's three children, James, Fonsiba, and Lucas Beauchamp, only Lucas remained. What was more, the blacks Isaac had seen during Reconstruction were no more capable of freedom than Bayard Sartoris's River Jordan marchers had been. Isaac had had bitter experiences with them. His black cousin Fonsiba had married an impressive, literate northern black who seemed to know what he was doing. America was going to be a "new Caanan," he had insisted. "We are seeing a new era, an era dedicated, as our founders intended it, to freedom, liberty and equality for all." He had taken Fonsiba to Arkansas, where he had a farm; but when Isaac visited them he was appalled. That farm was "a farm only in embryo," Isaac saw; it might even be "a good farm" someday, "but . . . not for years yet and only then with . . . unflagging work and sacrifice." There, in a ramshackle cabin, Fonsiba's husband sat reading a book through lensless glasses. "The curse you whites brought onto this land," he insisted blandly, "has been voided and discharged." To the horrified Isaac, Fonsiba's husband seemed to exude "from his skin itself, that rank stink of baseless and imbecile delusion" (*GDM*, 279, 277, 278).

How could such a people be free? It was here, paradoxically, that Isaac's notions of black behavior and his theories about God's curse merged into an apocalyptic vision of human liberation. Isaac was the first of Faulkner's protagonists to see paternalism as a self-fulfilling prophecy. Blacks' vices were "vices . . . that white men and bondage have taught them," Isaac insisted, and blacks could escape from God's curse only when white men escaped it. Isaac knew now that whites could "not resist it, not combat it"—nobody could; but because "my people brought the curse onto the land," he thought, "maybe for that reason their descendants alone can . . . just endure and outlast it until the curse is lifted." Then, he told Fonsiba's husband, blacks might finally have their freedom: "Then your peoples' turn will come because we have forfeited ours." That was why, instead of trying to save blacks from themselves by perpetuating the plantation, whites had to do something else: they had to free the black man's spirit by freeing their own and those of their children. Blacks, however, could take hope because they were "better than we are" and "stronger than we are"; for Isaac, they would "endure. They will outlast us" (*GDM*, 279, 277, 278, 294).[13] And that was why, on his twenty-first birthday in 1888, old Carothers's grandson renounced any claim to the land his father had bequeathed him.

III

To Faulkner, *Go Down, Moses* must have come through as the inevitable visionary extension of *Absalom, Absalom!* and *The Unvanquished.* Carothers McCaslin brought together all aspects of the curse Faulkner has visualized in Sutpen; Cass Edmonds, for all his failures, was the most articulate spokesman for the Cavalier ideal Faulkner had yet conceived; and in Isaac McCaslin Faulkner seemed to have discovered a protagonist who was the spiritual antithesis of Quentin Compson, a man who could burrow under both Puritan and Cavalier heritages to discover the possibilities for change. At this stage of his career Faulkner was not to be satisfied with a simplistic resolution of the problems that had destroyed Quentin; the woman Isaac married, shocked when she realized he was not going to reclaim the plantation, denied him the child for whom he claimed to have made his gesture. There were other ironies, too. Isaac's stand meant that the one McCaslin who understood God's meaning for the land had turned it over to those who did not. It meant that Cass's son Zack and his grandson Roth would inherit the land, and, as Isaac stuck stubbornly with his word, every failure Roth was responsible for seemed to point at Isaac. God's method might be evolutionary; but Isaac had not fully understood the terms of that evolution, and by the time he was in his eighties he was a tragic old man who seemed out of touch with God's plan.

Faulkner took care to show his readers what his protagonist could not see. *Go Down, Moses* swarmed with images of evolution and change. Quentin, ritually repeating that he did not hate the South, had been overcome with the prospect that the Burdens had feared: that the curse planters invoked would work through miscegenation to bring about an inferior species, presaging "the end of this nation's life." Rosa Coldfield had thought Clytie Sutpen's blood contained inimical strains that were somehow at war with each other; that was why "in the very pigmentation of her flesh" Clytie represented to Miss Rosa the tragic legacy of the war. Cass Edmonds saw Sam Fathers in somewhat the same terms. Sam's part-white slave mother "had bequeathed him not only the blood of slaves but even a little of the very blood which had enslaved it," Cass thought. To Cass, Sam, like Clytie, was "himself his own battleground, the scene of his own vanquishment and the mausoleum of his defeat (*GDM*, 168).

But *Go Down, Moses* sounded another note, too. Horses shied from Old Ben, but Isaac's one-eyed mule was a reliable mount because of his disability. Lion, the big dog who signaled the end of the wilderness by baying the old bear, was a mongrel. So was Isaac's "nameless and mongrel and many-fathered" little fyce, from whom he claimed to have learned "humility and pride" (*GDM*, 296). The fyce, Faulkner said later, "represents the creature who has coped with environment" because, "instead of sticking to his breeding and becoming a decadent degenerate creature," he had "mixed himself up with the good stock where he picked and chose."[14]

The ability of animals to adapt for survival had its corollary in humans and suggested the futility of Isaac's fears. It was Sam's white blood, Cass told Isaac, that caused the trouble; primitive African and Indian bloods fused harmoniously.[15] Lucas Beauchamp, Isaac's black cousin who was Turl's son, fared even better. His primitive heritage was as obvious as Clytie's and Sam's; Roth Edmonds, looking into Lucas's face, thought he saw "a man most of whose blood was pure ten thousand years when my own anonymous beginnings became mixed enough to produce me." For Roth, whose white blood came from the same source as Lucas's, the concession was significant: "pure" primitive blood was stronger than "mixed" white blood. It explained, perhaps, why Lucas had never experienced the blood-war from which Sam suffered. "Instead of being at once the battleground and victim of the two strains," Lucas "was a vessel . . . in which the toxin and its anti stalemated one another." For Roth, as for Cass, there was no question but that the white and black strains ("the toxin and its anti") were inimical; but Lucas's blood seemed to represent a rare, benign combination. That, Roth decided, was because of the strength of Lucas's "pure" African heritage. "It was as if he was not only impervious to . . . [his white] blood, he was indifferent to it"; Lucas "resisted . . . [the white blood] simply by being the composite of the two races which made him, simply by possessing it" (*GDM*, 71, 104).

For Isaac, such considerations made it all too easy to believe that he himself had inherited the worst of both worlds. Sam and Lucas had been able to mix themselves, like the fyce, "with the good stock where . . . [they] picked and chose," but Isaac had only his white blood—blood, Roth thought, that was hopelessly "mixed," that came from "anonymous beginnings," and that had no "pure" primitive blood to leaven it. And yet, hopelessly alienated from his family because of the primitive beliefs he had learned from Sam, Isaac was a product of a kind of spiritual miscegenation that had led him to the same cul-de-sac as Sam; like Sam, he was an anomaly in a society that had forgotten the life-style of the wilderness. That was why Isaac, like Sam, had no son; he was no more than "his own battleground, the scene of his own vanquishment and the mausoleum of his defeat." But although Isaac was a spiritual half-breed, he had also inherited certain benign qualities from that mixture.

As the wilderness slowly attenuated, Isaac, nearing eighty, still pursued the ritual of the autumn camp; he and the other hunters were forced now to drive far into the Mississippi Delta to reach the wilderness. But an aging Isaac seemed surprisingly vulnerable, clinging blindly now to his optimism about God's plan. "I still believe," he told the misanthropic Roth. "I see proof everywhere." For Isaac, that "proof" was human beings. "There are good men everywhere, at all times," he insisted, and he was sure that "most men are a little better than their circumstances give them a chance to be." Roth, who was showing the moral attenuation the curse brought, found the old man's optimism maddening and lashed out at him. "So you've lived almost eighty years," he said. "And that's what you finally learned about the other animals you lived among." He added angrily, "I suppose the question to ask you is, where have you been all the time you were dead?" (*GDM,* 347, 345).

Roth's bitter words showed Isaac, as nothing else could have, the moral consequences of his gesture; by turning the plantation over to Roth's grandfather, Isaac had assured that he himself would have a stake in Roth's degeneracy. But that insult was the least of the shocks Isaac was destined to have at this camp. Later, Roth handed Isaac an envelope of money and asked him to give it to a woman who would come while the younger men were hunting, with the message that Roth would not see her again. Roth did not know it, but that encounter would call Isaac's whole life into question. Roth's mistress understood God's curse as well as Isaac did, and she blamed Isaac for what had happened to Roth. "I would have made a man of him. . . . You spoiled him," she said, when you "gave to his grandfather that land which didn't belong to him." What was worse, the woman had a baby with her, obviously Roth's; and then, at last, something else came together in Isaac's mind. "You're a nigger!" he gasped, and with a flash of insight he saw what Roth had never seen: she was their cousin. "Yes," the girl told him. "James Beauchamp . . . was my grandfather" (*GDM,* 360, 361).

Isaac's serenity was shattered. "Get out of here!" he cried hysterically. "Cant nobody do nothing for you!"—and in his rage it suddenly seemed to him that the curse he had struggled all his life to elude was all too close to him. "Maybe in a thousand or two thousand years in America," he told himself. "But not now! Not now!" He found that his mouth was "running away with him," that he was unleashing a tirade that seemed strange on the lips of the man who thought blacks were "better" and "stronger than we are." "Marry a man in your own race!" he pontificated to his near-white cousin. She was "young, handsome, almost white"; he was sure she "could find a black man who would see in you what it was you saw in . . . [Roth]." And he concluded, "That's the only salvation for you—for a while yet, maybe a long while yet. We will have to wait" (*GDM,* 361, 363).

For Isaac, this reversal was tragic: the man who had given up everything to free his hoped-for son from God's curse had finally cried "nigger," and now he found himself rejecting his own cousin and her child. Isaac's beliefs about miscegenation had never been any different from Cass's and Roth's. All mulattos, he seemed to be saying, were tragic. Miscegenation created a war between their black and white bloods, and, torn by the desire to reject their black heritage and regain their white, all of them would remain tragically alienated; that was why Roth's mistress had gravitated unerringly toward her white family and why she was probably destined for further tragedy at their hands. But Isaac's diatribe was more than a senile lapse into an ancestral racism. He did not mean black was not beautiful; he was saying mulattos did not think so. And his outrage was not directed at the girl's primitive blood; that was pure and hence benign. For Isaac, that was the point. Years ago he had told Fonsiba's husband that his own people would have to "endure and outlast" God's curse and that only "then your peoples' turn will come." The source of Isaac's anguish was that, since Sam had marked him with the blood of his first kill, he had identified himself as one of those who like Sam had been dispossessed of the land. What Roth's mistress had shown him was that the black man's "turn" was going to be delayed still longer, and his anguished response only showed what that meant for his own life. "*We* will have to wait," he told the girl.[16]

Isaac should have looked more closely. Racial mixture, as Faulkner had shown throughout *Go Down, Moses,* signaled change; but it was never a signal of disaster, and a younger Isaac, struggling to free his heir from God's curse, might have seen more. Roth's mistress was a superior individual: a strong-willed, dedicated woman who had come back to the South not because she wanted to rejoin the McCaslins, but because, as a teacher, she was dedicated to helping fellow blacks achieve a better life. As her comments about Roth showed, she knew the meaning of God's curse. She had taken the responsibility of not telling Roth they were cousins, and she had borne the child knowing that because she was black Roth would never marry her. Like Isaac, she had learned how to deal with aloneness, comporting herself with a self-confident dignity. And perhaps because of her primitive heritage, she was capable, as few characters in the novel were, of giving love as well as receiving it. "Old man," she said when Isaac told her to marry a dark-complected man, "have you lived so long and forgotten so much that you dont remember anything you ever knew . . . about love?" (*GDM,* 363). Blundering Roth had stumbled on the woman Isaac should have had, and it was fitting when, in a moment of inspiration, Isaac gave her child one of the last souvenirs of his youthful experiences in the wilderness: the hunting horn that old

General Compson had given him. This youngest McCaslin, it appeared, might be Isaac's spiritual heir.[17]

But for Isaac it was too late; his gesture was one of blind hope in the face of total disaster. When the woman and her child were gone, he lay on his cot like a tomb effigy, "the blanket huddled to his chin and his hands crossed on his breast," agonizing over what the curse his grandfather had brought onto the land had done to its peoples. In "The Bear," when General Compson had tried to stop the ravaging of the wilderness by making part of it a private hunting club, Isaac had sounded like a young Thomas Dixon, Jr., complaining about miscegenation; one could not "change the leopard's spots," he insisted, if one "could not alter the leopard."[18] Now, in the Delta, "deswamped and denuded and derivered in two generations," he saw the completion of that process: the spots were still there, and they were growing. It was not whites alone who were guilty of destroying the old purity; this Eden had been destroyed not only "so that white men can own plantations and commute every night to Memphis," he reflected, but also so that blacks could own them "and ride in jim crow cars to Chicago to live in millionaires' mansions." And blacks were not the only victims; here "niggers crop on shares and live like animals," he reflected, but here also "white men rent farms and live like niggers." The land remained unbelievably fertile—cotton grew "man-tall in the very cracks of the sidewalks"; but human greed was just as fertile—that cotton grew amid "usury and mortgage and bankruptcy and measureless wealth." Together, the greed and fertility insured the absolute dissolution of the old purity, and the hallmark of its destruction was a universal miscegenation: in the ravaged Delta, Isaac reflected, "Chinese and African and Aryan and Jew, all breed and spawn together until no man has time to say which one is which nor cares." For Isaac, that was a vision as apocalyptic as Dixon's perception of a mulatto America, and he was devastated. "No wonder the ruined woods I used to know dont cry for retribution!" he told himself. "The people who have destroyed it will accomplish its revenge" (*GDM*, 363, 315, 364).

That was the final tragedy of Isaac's struggle with God's curse. Wedded to an archaic "purity," this most perceptive of Faulkner's old-generation protagonists was blind to the benign work of change all around him. Because of his blindness he could see no further than Quentin had; Isaac's final vision showed him only a decadent heritage, ravaged by a universal miscegenation, its future a genetic catastrophe. But Roth's child, as "Delta Autumn" showed, was not going to be any Jim Bond. His mother, more even than Lucas, was "a vessel . . . in which the toxin and its anti stalemated one another." The last black McCaslin, Faulkner seemed to be saying, might be a benign mixture of the old primitive stock with the better strains of a decadent white one: a new man who like Isaac's fyce had "coped with his environment" because he had "mixed himself up with the good stock where he picked and chose." And if that had happened on the black side of the McCaslin family, Isaac himself was evidence that it might happen on the white side, too. Isaac's life ended tragically; but his youthful gesture of repudiation remained to represent hope for the future. One aristocrat, however limited his insight, had learned ways, if not to transcend God's curse, at least in some measure to "endure and outlast it"; if there had been one Isaac, perhaps there might be others.

IV

As Faulkner looked back on *Go Down, Moses,* it must have seemed to him that the abolitionist South he had discovered in *A Dark House* had taken him a long way from the bland propaganda of the world of his youth. By 1942 the masking rituals of the Sartoris stories must have seemed a part of a distant past. He had had the courage to look behind those masks, he thought, in *Absalom, Absalom!,* and he had struggled in that novel, as few had before him, with the curse of inherited guilt. Faulkner's new novel seemed to offer, if anything, a more direct approach to Clytie's secret than *Absalom, Absalom!* Countrified, ignorant Carothers McCaslin was grosser, more barbarous even than redneck Thomas Sutpen; the McCaslin plantation had been, in Isaac's words, "founded upon injustice and erected by ruthless rapacity"; and the methods of that more knowledgeable theorist of McCaslin paternalism, Cass Edmonds, amounted to "downright savagery" (*GDM,* 298). Faulkner seemed ready to face the worst of Quentin Compson's fears: to examine the prospect that the patriarchs of even the best plantation families had sometimes been more like Sutpen than John Sartoris. He thus conceded the abolitionists a formidable part of their criticism of his heritage and indicated that the only hope for southerners might have been in extreme gestures such as Isaac's.

It was on that gesture, finally, that Faulkner's new novel was predicated. Isaac's life had ended tragically; but Faulkner made his gesture of repudiation stand for a turning point in the white South's long struggle with its conscience. It afforded a hope, however faint, for freedom from the curse that had held black and white southerners in bondage, and that meant that, for Faulkner, *Go Down, Moses* was more important even than *Absalom, Absalom!* Faulkner's new novel pointed directly toward the place he thought his search for a South was taking him. It foreshadowed a South in which others might succeed where Isaac had failed, a South in which the values—both primitive and paternalistic—which Isaac had stood for, might be marshalled to redeem the patrimony Faulkner's protagonist could only separate himself from. That, finally, was why Faulkner had been at such pains to make the secrets of the McCaslin commissary more Gothic even than Clytie's; and it was why he had been at such pains to create a protagonist who would not flinch, as Quentin had, when he confronted those secrets.

It was on that commissary debate between Cass and Isaac that the meaning of these secrets hinged. If Cass was right and blacks were children who had to be protected from themselves, the plantation, despite its shortcomings, was justified as an institution. If Isaac was right and blacks were "better" and "stronger" than whites, the plantation was a travesty, and Isaac's repudiation was the only way out. Intensely aware of that dichotomy, Faulkner had populated his novel with blacks who seemed to illustrate one or the other of these ideas. *Go Down, Moses* presented a panorama of the effects of slavery and manumission on five generations of McCaslin slaves; and these effects were particularized in two part-black characters—Sam Fathers and Lucas Beauchamp—and in two who had no significant white ancestry—Rider (a sawmill worker who lived on the plantation[19]) and Lucas's wife Molly.

These figures isolated the problem Faulkner was facing as he struggled to create a hero who could transcend his heritage. The only black Isaac was shown to have had much contact with before the commissary dialogue was Sam Fathers. Socially, Sam was black; but Faulkner showed the reader hardly anything about Sam's contact with blacks. His part-black mother, like Clytie Sutpen's, was given little attention, and even Isaac had trouble seeing African physical characteristics in Sam; the "only . . . trace of negro blood" Isaac could observe "was a slight dullness of the hair and the fingernails." Faulkner was better with the psychological scars of slavery, which were more apparent; there was "something . . . about the eyes," Isaac sensed, "which you noticed . . . because it was not always there," and it was "not in their shape nor pigment but in their expression." Edmonds explained: it was the "mark . . . of bondage," the "knowledge that for a while . . . part of his blood had been the blood of slaves" (*GDM*, 167).

The thinness of the African side of Sam's characterization revealed the problem. If Sam identified himself wholly with his Indian ancestors, ignoring his African and white ones, it was a telling comment on his status as a slave; but it did next to nothing to buttress Isaac's theories of black virtue. "Pride" and "endurance" might indeed be black virtues, but Sam, it appeared, had learned these from Indians, not blacks. Sam was no typical slave. It was a rare slave who enjoyed the privileges of working when he pleased and of addressing his masters — as Sam did — as equals. Faulkner had provided Sam with little to show for his African ancestry and given him no dramatized understanding of it. Isaac claimed that Sam had taught him "the long chronicle of a people who had learned humility through suffering and learned pride through the endurance which survived the suffering" (*GDM*, 205); but the only blacks with whom Isaac was shown in direct contact before the debate with Cass were Fonsiba's husband and Sam.

The situation was confusing, and the confusion multiplied when Faulkner brought in other blacks, with whom Isaac had not been shown in contact before the debate, to illustrate what he had failed to show in Sam. The most obvious of these was Lucas Beauchamp, old Carothers's black grandson, who appeared in the novel more often than any other character except Isaac. Again, Faulkner drew on familiar memories, this time of Ned Barnett who, except for the lack of blood ties, had somewhat the same relationship to Faulkner that Lucas had had to the Edmondses. More than any other character in Faulkner's canon, Lucas seemed to resemble Ned. Faulkner had given Lucas a few of Ned's superficial qualities. A finicky dresser who copied the sartorial styles of Isaac's grandfather as Ned copied those of Faulkner's father and grandfather, Lucas also shared Ned's affinity for wanting "to take part in everything and boss all of it"; and in "The Fire and the Hearth," as Faulkner recounted Lucas's old-age problems with his wife Molly — who pointedly resembled Caroline Barr — Faulkner seemed to be describing nothing so much as a part-white Ned married to Mammy Callie. But Faulkner had more in mind here than another sentimental invocation of the world of his youth;[20] Lucas's portrait was one of his finest, and it promised a reinforcement of Isaac's ideas that would compensate for Faulkner's failure with Sam's.

The first impression that "The Fire and the Hearth" gave, however, made Lucas no more than another black clown. Readers were apparently meant to laugh at his way of speaking. A divorce was a "voce" for Lucas, and a bill of sale, a "billy sale." Lucas was never shy about playing darky; trying to pin him down, Roth Edmonds found, was impossible, for the old man "became not Negro but nigger" — he became "not secret so much as impenetrable, not servile and not effacing, but enveloping himself in an aura of timeless and stupid impassivity almost like a smell" (*GDM*, 128, 93, 60). But framed by that comic Lucas was a portrait of a younger, more heroic man. Forty-three years before the gold hunt, Roth Edmonds's mother had died giving birth to Roth; Molly, nursing Lucas's first child, had delivered that white baby and stayed at the home of Roth's father Zack to nurse both infants. Nearly half a year passed, and

Lucas began to see the situation as a crisis of his self-respect. "I got to kill him or I got to leave here," he concluded, and went to Zack's house. "I'm a nigger," he told Zack. "But I'm a man too." After some verbal sparring, he seized Zack's pistol and pulled the trigger. It misfired, but the incident had meant everything to Lucas. "I would have paid," he reflected proudly. "I would have waited for the rope, even the coal oil" (*GDM*, 49, 47, 58).

Lucas had defended his manhood as few blacks in his time were able to, and now he exuded a mature self-respect. His strength reinforced Isaac's claim that blacks are "better" and "stronger" than whites; and Lucas himself seemed to illustrate Isaac's belief that blacks had "learned pride through the endurance which survived the suffering" of slavery. The trouble with Lucas was that his presence, like Sam's, created difficulties of which Faulkner seemed unaware. If Lucas's "suffering" as a black had taught him any "humility," it was not visible, nor did this arrogant old man show any signs of "pity and tolerance and forebearance." Even less visible was any dramatic evidence of that primitive purity Roth Edmonds thought he saw in Lucas. Lucas might be "indifferent" to his white blood, but, like Sam, he never thought about himself as a black; he thought of himself as an aristocrat deprived of his heritage by the accident of black ancestry.

More than any white McCaslin, Lucas patterned himself after old Carothers. His pride in his heritage from this ancestor amounted almost to a phobia. The reason Zack double-crossed him about Molly, Lucas insisted, was "because what you and your pa got from old Carothers had to come to you through a woman—a critter not responsible like men are responsible." Even while Lucas was insisting, "I'm a nigger. . . . But I'm a man too," he was doing something more than affirming black manhood. "I'm more than just a man," he told Zack. "The same thing made my pappy that made your grandmaw." The very mark of his courage in confronting Zack was his willingness to repudiate that heritage. Lucas's white blood must not be "worth much," he told Zack, "since old Carothers never seemed to miss much of what he gave to Tomey"; maybe it "rightfully aint even mine," he admitted; he insisted, "If this is what that McCaslin blood has brought me, I dont want it neither." But after Lucas had proved his courage by challenging Zack, his real feelings showed themselves. "I reckon I aint got old Carothers' blood for nothing," he reflected. "I needed him and he come and spoke for me." The effects of God's curse were subtle. With Sam Fathers, Lucas stood as an illustration of whatever benign effects racial mixture might produce; but if Lucas, like the fyce, had "mixed himself up with the good stock where he picked and chose," he at least seemed to think that the good stock came entirely from Carothers McCaslin. In believing this he was not alone. Lucas, Roth concluded, was "more like old Carothers than all the rest of us put together, including old Carothers" (*GDM*, 52, 47, 57, 58, 118).

Lucas's affinity with his white ancestor showed the pathos of the curse Carothers had brought onto the land: by committing his blood to slaves, Carothers had disinherited his spiritual heir and provided his white descendants with a living symbol of their guilt. But subtle as Lucas's characterization was, it suffered from unresolved ambiguities. If Lucas had "learned pride through the endurance which survived the suffering," his portrait argued otherwise; his identification with his white grandfather implied that Lucas's pride came instead from this ancestor. Lucas's very strength created an unintentional irony, for if old Carothers's spiritual heir had inherited the strength to ward off the effects of God's curse, the curse could then have little retributive value. Lucas's portrait, like Sam's, undermined Faulkner's efforts in the novel as much as it facilitated them.

Lucas and Sam were counterpoised in *Go Down, Moses* with two complex portraits of blacks who had no identifiable white blood[21]: Lucas's wife, Molly, and Rider, a sawmill worker who lived on the plantation. "Pantaloon in Black" related how Rider was lynched for murdering a white gambler. It was one of Faulkner's most ambitious tales. Superficially, it seemed no more than a retelling of earlier lynching tales like "Dry September" and *Light in August*, but Faulkner was still experimenting. Unlike Joe Christmas, Rider was a black raised among blacks; unlike Will Mayes, Rider received a full portrait. "Pantaloon in Black" was the only story Faulkner wrote in which every significant character was black; it was only after Rider was dead that Faulkner introduced a white deputy and his racist wife to provide further perspective. Determined to show what was happening to Rider from within, Faulkner entered Rider's mind for the only lengthy stream of consciousness he ever attempted with a full-blooded black. Finally, Faulkner intended Rider as a dominant black hero, the only such portrait in his canon. Rider was a magnificent laborer of "midnight-colored" skin, "better than six feet," and weighing "better than two hundred pounds"; he worked in a lumber crew; sometimes "out of the vanity of his own strength" he would life "logs which ordinarily two men would have handled with canthooks." He was a natural leader; at twenty-four he was the "head of the timber gang itself because the gang he headed moved a third again as much timber . . . as any other moved" (*GDM*, 144, 135, 137). An orphan raised by a devout aunt and uncle, he had just made a good marriage. For Rider, at least, life on the McCaslin plantation seemed rewarding; but after six happy months Mannie, his bride, was dead.

Faulkner conceived a shrewd strategy with Mannie: he never revealed how she died. What interested Faulkner was what the tragedy did to Rider: in a strange outburst of emotion, Rider cut all ties to his former life, including, pointedly, his religion. Sleepless, he rambled the countryside, drinking, getting more and more hysterical, ranting at God in a drunken tirade. Finally, for no apparent reason, he murdered a white gambler, Birdsong, and a little later he was found quietly asleep on his front porch. In jail his hysteria returned, and he became so violent that the other prisoners had to pin him to the floor. Soon Rider got what he apparently expected: Birdsong's relatives took him from jail and hanged him. The striking thing about all this was how little the things that had destroyed Rider seemed to grow out of his relationship with whites. For Faulkner, that was the point: in spite of fate and Rider's own character, his tragedy still remained racial. That was why Faulkner never told how Mannie died; giving *any* reason for her death would have connected Rider's grief to the accidents of black life. As it was, Rider could account for it only as an act of God, and that pointed directly at the roots of his hysteria: in his heart, Rider believed all black tragedies came from whites. Those were his deep, permanent feelings, and they ignored the logic of the situation, denying him rest, until they found a racial outlet in his killing of Birdsong.

But the same problems that Faulkner had experienced with Sam and Lucas surfaced in Rider's story as well. What happened when Faulkner attempted to enter the mind of his black hero was typical; much of what went on in Rider's mind seemed to have little to do with a nonliterary black's way of looking at himself. Rider's return to the house in which his love was consummated, his encounter with Mannie's ghost, his attempt to drown his sorrows in alcohol, even his longing for a death he finally achieved—all of this was as old as Gothic romance and said nothing unique about blacks. Moreover, what did come through as black about Rider was largely superficial, an almost mechanical repetition of surface features of Mississippi black experience. Too often these were near-clichés. Rider carried a razor and drank whiskey from a jug; he and Mannie ate sidemeat and greens and cornbread and drank buttermilk; he called whites, singular and plural, "white folks"; he punctuated a crap game with cries of "Ah'm snakebit" (*GDM*, 147, 152–53).

Rider's behavior did little for the credibility of Isaac's beatitude of black virtues. Rider might be "better" than whites spiritually; his emotions, like his physique, might be "stronger"; with Mannie, at least, he showed "fidelity" and was no stranger to "suffering." But Rider's experiences taught him no "humility," and his soaring pride was hardly the kind that Isaac thought developed "through the endurance which survived the suffering." Indeed, Rider himself did not endure. If Rider was an example, "endurance" was not for black males, however heroic; it was reserved in *Go Down, Moses* for black mammies and men descended from whites and Indians.

The burden of the task of illustrating Isaac's beliefs fell, finally, on Molly Beauchamp, and this seemed fitting. Faulkner had dedicated his novel to Mammy Caroline Barr, and Molly, more than any other character he had hitherto created, was patterned on Mammy. As with Lucas and Ned, there were surface similarities. Mammy was a "small" person, John Faulkner wrote, who "weighed only ninety-eight pounds"; Molly, Roth Edmonds observed, was "a small woman, almost tiny." Mammy, John wrote, was usually decked out in a "head rag" and a "fresh-starched dress and apron;[22] Molly always appeared in a "clean white headcloth and aprons" (*GDM*, 100). But more important, as Roth struggled to articulate his feelings about Molly, they sounded pointedly like the feelings Faulkner had expressed for Mammy Callie in his dedication. Molly, Roth reflected, had served him "without stint or expectation of reward"; Mammy, Faulkner wrote, "gave to my family a fidelity without stint or calculation of recompense." Molly, Roth said, had provided a "constant and abiding devotion and love" (*GDM*, 117); Mammy, Faulkner wrote, had given "to my childhood an immeasurable devotion and love."

Though Molly was never shown with Isaac, her presence pervaded *Go Down, Moses*. Her two chief appearances seemed to reveal the very essence of what paternalists claimed to stand for: each was an emotional appeal to a white aristocrat for help in maintaining the proprieties of her home. In "The Fire and the Hearth," the novel's second tale, Molly appeared in the commissary to ask a shocked Roth to get her a divorce from Lucas, whose treasure-mania she thought was bringing the curse on their home. In the closing story, which bore the novel's title, "Go Down, Moses," Molly appeared in County Attorney Gavin Stevens's office to plead for help in finding her lost grandson Butch. The significance Faulkner placed on Molly could be seen in the way he handled those appeals. He had given her something of the same treatment he had given Dilsey and Clytie; an important minor figure in the earlier story, she emerged in "Go Down, Moses" as the novel's tragic heroine. But Faulkner was attempting something different with Molly, too; as devoted as either Dilsey or Clytie to her white family, Molly was also hopelessly dependent on her whites, and her tragic life emerged as a symbol of the meaning of God's curse for the McCaslins' paternalism.

If Sam, Lucas, and Rider were poor symbols of Isaac's black virtues, Molly reflected all of these virtues effectively. That tragic life had been predicated on "pity and tolerance and forbearance and fidelity and love of children." She had "learned humility through suffering and learned pride through the endurance which survived the suffering"; Molly was something of a miracle of "endurance." Her fearless morality argued for Isaac's belief that blacks "are better than we are" and "stronger than we are." And if Molly's sanity had been destroyed by the time she accused Roth of selling Butch to Pharoah, the primitive faith that focused her grief came through as evidence of a mysticism similar to what Isaac had inherited from Sam.

Still, the problems Faulkner had encountered with Sam, Lucas, and Rider remained. If Molly illustrated Isaac's black virtues, it was also true that she lacked the one virtue blacks prized most: independence. Molly never appealed to other blacks when she was in trouble; she turned to whites like Edmonds and Stevens. That reaction introduced a fatal element into the novel: a degree of sentimentality that he had not allowed himself since the last portrait he had based directly on Caroline Barr, Callie Nelson in *Soldiers' Pay.* Faulkner showed Molly almost entirely through the eyes of those two pillars of the aristocratic community, Gavin Stevens and Roth Edmonds; and they responded to her pleas as though they were knights responding to the call of a lady in distress, driving themselves frantically to satisfy her simplistic desires as though her life could be made worthwhile only through their efforts.

Faulkner was clever with Stevens and Roth; their guilt testified to the reality of the curse Isaac had struggled to escape. The trouble was Molly; if she was the novel's most effective image of Isaac's black virtues, she also illustrated their limits with cruel precision. Molly believed without question that these knightly paternalists were there to fulfill her needs. To Stevens she said blandly, "You the Law. I wants to find my boy." Confusing Roth with his father on one occasion, she called him simply, "marster" (*GDM*, 371, 102). That Molly turned to Roth and Stevens rather than to blacks in her time of need revealed her as the epitome of the black child of plantation legend, and her presence in the novel's closing episode insured that its final image would be more than a tragic picture of the ravages of God's curse. *Go Down, Moses* closed with a stereotyped image that reflected the propaganda of the world of Faulkner's youth: the pathetic dependent and the obliging paternalist, bound together by sentiment.

Faulkner's black and mulatto portraits in *Go Down, Moses* served well as illustrations of Cass Edmonds's shallowness; but they did little to suggest the validity of Isaac's supposedly more perceptive views. In fact, the image of black life that emerged bore only an ambiguous resemblance to those views. The image of black manhood revealed through Sam Fathers and Lucas Beauchamp was fatally obscured by their identification with non-black ancestors, while that revealed through Rider suggested that black males were too physical, too emotional, and too childish to survive the rigors of Yoknapatawpha life. The image of black womanhood revealed through Molly was one from which all eroticism had been censored, which focused itself solely on domesticity, and which suggested black women were childish to the point of helplessness.

What was perhaps most striking about Isaac's list of black virtues, in fact, was the childishness which the virtues themselves implied. This childishness was the only point he and Cass agreed on, and their agreement suggested a great deal about both Isaac and the virtues he praised. "Endurance," "fidelity," "humility," "pity," "tolerance," "forbearance," and even "love of children"—every quality Isaac listed except "pride"—were not the virtues of independence. Isaac's beatitudes were no romantic appeal to a primitive strength; they were, instead, a primer of what aristocrats expected of a serviceable slave, and even the "pride" Isaac boasted of had been "learned through the endurance which survived the suffering." *Go Down, Moses* was not a book about how the aristocratic tradition might be renewed through contact with the primitives that the aristocrats had disinherited. If anything, it was a book about how one aristocrat could renew himself without probing very far outside the limitations he had inherited from the world of his youth.

Still, Faulkner's final monument to Mammie Callie was a remarkable achievement. His farm had yielded a crop richer than any that he or his tenants had produced; living the life of "farmland and horses and the raising of grain and feed" and experimenting with tenants who farmed on an agreement of mutual trust, Faulkner had been able to explore firsthand his notion that "land did not belong to the people but . . . people belonged to the land." Probing that belief may have enabled Faulkner to escape in *Go Down, Moses* many of the snares that had plagued him in the past. Old Carothers's crimes against the land and its people had been outrageous, and Faulkner had not used the McCaslin patriarch's un-aristocratic background to divert the guilt for those crimes from the aristocratic class; the patrician Edmondses had inherited old Carothers's guilt with his property and hence were destined to continue his outrages against the land and its people. In *Go Down, Moses* Faulkner seemed determined to allow no aristocrat, however formidable his pedigree, to avoid the suffering caused by the curse of inherited guilt. And in Isaac McCaslin Faulkner had created an aristocrat who was capable of learning, through contact with the other races who inhabited the land, how one individual might escape that curse in some measure.

But if Faulkner's white characters in *Go Down, Moses* suggested a more open approach to the South's guilty past, his black characters did not. Superficially inserted to reveal the motivation for Isaac's repudiation of his heritage, they revealed, instead, that his repudiation was founded on quicksand; for if Isaac had not understood the primitives who prompted him to the gesture, he had not understood the gesture itself. That fact boded ill for the future of Faulkner's search for a South. *Go Down, Moses,* Faulkner had hoped, would point the way toward a visionary South where the values Isaac had stood for might redeem the wasteland Quentin had been unable to survive. But if this most soul-searching of Faulkner's efforts to understand the world of his youth had foundered on the same reefs its predecessor had foundered on, his future efforts were not likely to fare better.

10

A Once and Future South

I

Aristocrats, the popular stereotype ran, were people who owned land and white-columned houses, and William Faulkner now owned both; but by 1940 the paternalist image was wearing thin. Faulkner had "become the sole principal and partial support," he wrote Robert Haas at Random House, "of my mother, . . . [a] brother's widow and child, a wife of my own and two step children, [and] my own child"; he had also "inherited my father's debts and his dependents, white and black"; and for this formidable crew Faulkner had to come up with money for "food, shelter, heat, clothes, medicine, kotex, school fees, toilet paper and picture shows."[1] That, in part, was why Faulkner practiced his paternalism at long range much of the time, grinding out a meager salary at Warner Brothers Studio in Hollywood. He was worried about his integrity in that wasteland, he wrote agent Harold Ober; but as he struggled to express this fear he seemed to be talking about something much deeper. Writers who reached their "early fifties" were facing a crucial "period . . . which most artists seem to reach," he said, a time when "they admit at last that there is no solution to life and that it is not, and perhaps never was, worth the living."[2]

One "solution to life" everybody else seemed to be discovering was World War II; brothers John and Jack, nephew Jimmy, and stepson Malcolm Franklin were soon in uniform; but forty-five-year-old Faulkner had not been able to get a commission. As he wrote Malcolm in 1942, he seemed to be struggling to find a "solution" in which age would count for something. "We will have to make the liberty sure first, in the field," he said. "Then perhaps the time of the older men will come." He meant men who would not yet be "so old that we too have become another batch of decrepit old men looking stubbornly backward at a point 25 or 50 years in the past"; they would be "the ones like me who are articulate in the national voice, who are too old to be soldiers, but are old enough and have been vocal long enough to be listened to."[3] The novel Faulkner was working on was set in France in World War I. "The argument," he wrote Haas, was that "in the middle of that war, Christ . . . reappeared and was crucified again." Faulkner thought he knew where that "argument" was taking him. "We did this in 1918," his book would say, but "in 1944 it not only MUST NOT happen again, it SHALL NOT HAPPEN AGAIN."[4]

By the time Faulkner finally published this book as *A Fable* (1954), his country was involved in still another war, a global Cold War that had erupted in full-scale fighting in Korea. By then it was clear that *A Fable*, whatever its setting, was no diversion from his search for a South, and that being "articulate in the national voice" was very much a part of that search, too. Exploring the abolitionist South in *Absalom, Absalom!* and *Go Down, Moses*, Faulkner had still seemed preoccupied with old threats to the world of his youth: threats by Puritan Yankees against aristocrats, threats by Puritan rednecks against aristocrats and blacks. But the war and his years in Hollywood had shown Faulkner other threats: threats by the modern wasteland Hollywood represented and threats by foreign dictatorships. It must have seemed to Faulkner that he was now in an enviable position to carve out a role for himself as a kind of cultural elder statesman. As a southerner, Faulkner had inherited an understanding of the traditional values that the Old South had claimed as its own; what was more, he was a special kind of southerner, one who should be able to speak out with a clear conscience because he had faced in *Absalom, Absalom!* and *Go Down, Moses* the full impact of the guilt that had undermined those values. And by 1948 Faulkner must have believed that there were compelling reasons for him to make known his views on the issues of his time.

When the 1948 Democratic platform called for a federal anti-lynching bill and an end to discrimination in hiring, many southerners began to get a familiar feeling; to such people it was beginning to look alarmingly like they were facing a new Reconstruction, and that was the one eventuality most calculated to stir up old fears. Southerners could no longer hope to secede from the Union, but they could and did secede from the Democratic party and form their own Dixiecrat party. Other difficulties were on the horizon, too. The Cold War made many Americans realize that if they did not resolve their racial problems, there were enemies who would exploit them. To Faulkner, the situation presented a paradox: for eighty-three years, the values he thought the Old South stood for had been sealed off from the rest of the United States because the Civil War had discredited them. But America was going to need those values now more than ever. If America was going to be saved, the South had to be saved, and that meant that southerners had to clean their own house, and fast. If they did not, the federal government would clean it for them, and that might bring on a conflagration in which those values would again be discredited. To Faulkner, that meant that the most important moral and political battles of his time were going to be fought in the South; and those battles, in turn, called for a new kind of protagonist in his search for a South. In 1877 L. Q. C. Lamar and his associates had stepped forth as representatives of the values of the Old Order, the Redeemers who would save Mississippi whites from Reconstruction; Faulkner's new heroes were going to be the Redeemers for this new Reconstruction, and they would point the way toward a South where the values of the Old Order might be recycled to serve a new time.

In February 1948, with President Harry Truman about to ask Congress for civil rights legislation, Faulkner wrote Ober about a novel he was writing.[5] "The story is a mystery-murder though the theme is more [the] relationship between Negro and white," he said. Its "premise" is "that the white people in the South . . . owe and must pay a responsibility to the Negro"—that they owed it, he said, "before the North or the govt. or anyone else."[6] Faulkner had been ruminating on this new novel since Howard Hawks had told him that he ought to write a "detective story" a few years earlier. "I've been thinking of a nigger in his cell," Faulkner had replied then, "trying to solve his crime"[7]—that seemed to have governed his choice; once he decided on a black man, he said later, "the character of . . . Lucas Beauchamp came along," and "he took charge of the story."[8]

Lucas seemed a promising choice; the crotchety, hard-headed old man who had told Zack Edmonds he was "a nigger" but "a man too" was a model of black independence. That independence had finally gotten Lucas in trouble. He had been found standing over a dead white man with a recently fired pistol; the dead man was Vinson Gowrie, whose father, Nub, was patriarch of a murderous hill clan. With Roth Edmonds in a New Orleans hospital, Lucas now turned for help to a white who bore no paternalistic or blood responsibility toward him, sixteen-year-old Chick Mallison. The components of Chick's character were familiar. This patrician youth had been born the same month as black Aleck Sander, son of the Mallisons' maid, and the two had been raised like brothers; an older black with white ancestry, Lucas himself, had initiated Chick into the mysteries of race; and an older white aristocrat, Chick's uncle Gavin Stevens, had grounded him in class virtues. But Faulkner had prepared a different kind of initiation for Chick than Bayard Sartoris and Isaac McCaslin had experienced. Aggressive, fun-loving Aleck seemed like another Ringo Strother, but Faulkner kept him in the background while Lucas and Gavin emerged as the prominent influences in Chick's life. Unlike retiring Sam Fathers, Lucas had come to grips with the world he lived in; and Heidelberg-and-Harvard-trained Gavin was far more sophisticated, socially and morally, than Granny Millard or Cass Edmonds. With these mentors, Chick was better prepared than either Bayard or Isaac had been to cope with postbellum southern problems; more important, Chick brought something to these problems that neither Bayard nor Isaac possessed.

Like Isaac, Chick had taken his aristocrat's heritage for granted since childhood; but Faulkner was ready to add a new dimension in the southern loyalties he ascribed to Chick. Bayard and Isaac had valued their family heritages, and Bayard, under Granny's tutelage, had felt a responsibility for the white as well as the black poor of Yoknapatawpha County; but none of Faulkner's aristocratic heroes had displayed any feeling of ethnic solidarity with other whites. Chick did. A part of his heritage as a southerner was "the . . . passions and hopes and convictions and ways of thinking and acting of a specific kind and even race." Chick projected that ethnic identification onto all southern whites, plebeian as well as patrician. He felt that "they were his and he was theirs"; he had "a fierce desire that they should be perfect"; with them he hoped "to be found worthy to present one united unbreakable front to the dark abyss," and he had a "furious almost instinctive desire to defend them from anyone anywhere" (*ID*, 151, 209, 194).

But Chick's ethnocentricity had been modified by an encounter with a member of the other ethnic group that inhabited his native land. Hunting on the McCaslin plantation Chick had fallen into an icy creek, and when he looked up from the water he had discovered Lucas watching him. Chick was immediately aware of something different about Lucas; his face was "inside a Negro's skin but . . . what looked out of it had no pigment at all, not even the white man's lack of it" (*ID*, 6–7). Chick did not know how to respond. His first reaction was dumb outrage; to his dismay, he realized that the only relationship Lucas was going to acknowledge between them was that between a man and a boy—Chick's white skin counted for nothing. The old man commanded Chick to follow him home, where he and his wife Molly wrapped the boy in a blanket, dried his clothes, and offered him a meal. Confused and angered by such kindness from blacks, Chick tried to establish his white role by offering the old couple the change in his pocket, and when he was refused the boy was immediately aware something had happened; it was, in fact, very much like what a youthful Roth Edmonds had experienced with a younger Lucas and Molly in *Go Down, Moses* when "the old curse of his fathers . . . descended to him." Now, Chick realized, "by that one irrevocable second" it had taken to extend the coins, "he was forever . . . too late" to save himself. Later, he would feel a "need . . . for reequalization, reaffirmation of his masculinity and his white blood" (*ID*, 15, 26). But Chick was no Roth Edmonds; he struggled manfully with his guilt, and, when time at last afforded him a kind of freedom, he had had an agonizing crash course in black culture.

Like most southern whites, Chick had undergone a lengthy indoctrination by the time he was sixteen. Some of his earlier experiences, familiar but not yet evaluated, sounded embarrassingly like the preoccupations of the author of *Soldiers' Pay* and *Sartoris.* To Chick, the mongrel dog he and Aleck took on the rabbit hunt was "a potlicker, a nigger dog"; he needed only "one glance to see [that it] had . . . a rapport with rabbits such as people said Negroes had with mules." Huddled under Lucas's blanket, Chick felt "enclosed completely . . . in that unmistakable odor of Negroes." At the table he thought he was getting "nigger food"—to Chick, that was "collard greens, a slice of sidemeat fried in flour, . . . heavy half-cooked biscuits, a glass of buttermilk"; blacks, he thought, had "elected this out of all eating" because "their palates and their metabolism" craved it. But now Chick was ready to re-think those impressions. Food like that was all blacks "had had a chance to learn to like," and he theorized that the odor was the result of "a passive acceptance . . . of the idea that being Negroes they were not supposed to . . . bathe often . . . , that in fact it was a little to be preferred that they did not" (*ID*, 5, 11, 13). Chick was learning to push through to the human beings behind those stereotypes, and he was ready for an experience that would take him a long way from his preoccupations with mules and "nigger" dogs. Since Lucas's arrest Chick had seen no blacks on the streets, but he had a strange feeling "of their constant . . . nearness: black men and women and children . . . just waiting, biding." Riding a country road, he finally understood that response: the deserted countryside was "a witness . . . of the deliberate turning as with one back of the whole dark people," in "one . . . invincible . . . repudiation, upon not a racial outrage but a human shame" (*ID*, 96–97). The black response showed that the white countrymen Chick had been ready "to defend . . . from anyone anywhere" were guilty of something terrible. It was even more appalling when he compared those whites to Lucas.

Unflappable, master of every situation, the Lucas of *Intruder in the Dust* was the most admirable model of black manhood Faulkner had yet created. He had long ago mastered the semantics of segregation. Anybody "who knew that part of the country," Chick recalled, knew "about the Negro who said 'ma'am' to women just as any white man did and who said 'sir' and 'mister' to you if you were white but who you knew was thinking neither." When Lucas stepped out of the police car at the Jefferson jail, Chick observed, he bore an "arrogant" look, but his demeanor was "calm and with no more of defiance in it than fear." In jail Lucas slept soundly. When Gavin and Chick visited him, he asked no favors, refused to tell them any more than he wanted to, and insisted on paying for their assistance. When an angry Gavin told Lucas that "if you just said mister to white people . . . , you might not be sitting here," the old man asked sardonically why he was supposed "to commence now," and he had a suggestion. Maybe, he said, "I can start off by saying mister to the folks that drags me out of here and builds a fire under me" (*ID,* 18, 44, 62).

Chick had been fortunate to learn about blacks from Lucas. Isaac McCaslin had learned from Sam Fathers that a black man could live a dignified, rewarding life by rejecting a partitioned society; but Chick had learned that a black man could achieve those goals in the midst of one. Lucas's very manliness identified Chick's problem. The old man was so courageous, and the people about to lynch him so cowardly, that Chick was about to lose faith in all those whites that he had been ready "to defend . . . from anyone anywhere." His education about race was about to enter another phase. Lucas had given him a crash course in what being black meant; the lynch mob was about to give him one in what prejudice meant. But here again, Chick had the advantage over Isaac McCaslin, for in Gavin he had a white mentor who could interpret his white ethnocentricity better than Cass Edmonds had been able to interpret Isaac's.

To show Chick how prejudice operated, Gavin predictably went outside their class; he chose Mr. Lilley, a shopkeeper who catered to the black trade. Lilley was no hill farmer like the Gowries, but his redneck attitudes were still showing. "He has nothing against what he calls niggers," Gavin told Chick. "If you ask him, he will probably tell you he likes them even better than some white folks he knows." By Lilley's own lights he was generous; for instance, he ignored the fact that his customers were always stealing small items. "All he requires," commented Gavin, "is that they act like niggers." To Lilley that also meant that sooner or later one of them was going to kill a white man; that, Gavin reported, was what "Mr. Lilley is probably convinced all Negroes want to do." When Lucas was accused of killing Vinson Gowrie, it only confirmed the man in his beliefs; now he expected a lynching as a matter of course. It was all going to be a game played "observing implicitly the rules," Gavin said, with "the nigger acting like a nigger and the white folks acting like white folks and no real hard feelings on either side." Gavin seemed to think prejudice like Mr. Lilley's began with his un-aristocratic background; what Mr. Lilley proved, Gavin said, was that "no man can cause more grief than that one clinging blindly to the vices of his ancestors" (*ID*, 48–49).

Mr. Lilley represented the average plebeian white; such people, Gavin said, would passively accept Lucas's lynching, leaving the dirty work to others. Like other southern towns, Jefferson had a hard core of hoodlums who "never really led mobs" but "were always the nucleus of them because of their mass availability." These men were capable only of a mass, rather than individual, identification, Gavin explained, and lynchings drew them. When the mob finally gathered, Chick saw that their faces had become "curiously identical in . . . their complete relinquishment of identity" (*ID*, 137). But mobs had one weakness, Gavin said: the very pliability that made people mobs also made them helpless without leaders. It would not be Jefferson, finally, that would be responsible for lynching Lucas; the misguided individualism to accomplish that had to come from outside.

Now Gavin was ready to educate Chick in the geopolitics of lynching. Yoknapatawpha County was divided between Anglo-Saxons who lived in the valleys and Scotsmen who lived in the hill district known as Beat Four. The valley men sounded like those described in *The Hamlet*: their names were "Littlejohn and Greenleaf and Armstead and Millingham and Bookwright."[9] Because they had "the broad rich easy land where a man can raise something he can sell openly in daylight," they were not dangerous; it was the Scotsmen whom people had to watch out for. The ancestors of the Beat Four Scotsmen had chosen hill country because it reminded them of their native highlands. Farming came hard there, and their interest turned elsewhere; after all, as Gavin said, "it really doesn't take a great deal of corn to run a still." Those Beat Four people were clannish; the Gowrie clan included a father and six brawling brothers, not to mention "cousins and inlaws covering a whole corner of the county." Now "intermarried" into other families, they were "not even . . . a simple clan or tribe"; they were more like "a race" or "a species," and, as far as Gavin was concerned, that race was alien. Beat Four's reputation made it, he thought, the Southern equivalent of Cicero, Illinois (*ID*, 149, 148, 35). It was from Beat Four, at any rate, that the drive to lynch Lucas would originate, and, for the moment at least, neither Gavin nor Chick had any doubts that it would succeed.

Chick's education in race had been traumatic: a crash course in black culture followed by another in white prejudice. But knowledge, Chick learned, carried with it responsibility. Chick was now the only white person in Jefferson who not only understood white prejudice but also understood the black pride that had made Lucas refuse to tell what he knew of the murder. Thus, what Chick faced was going to be even more traumatic: he had to learn how to forestall a lynching. Lucas now confided to Chick that it was not his pistol that had shot Vinson Gowrie; but when Chick asked whose pistol *had,* the old man refused to say. "Go out there and look at him," he said. The task seemed hopeless for anyone, much less for a sixteen-year-old boy; but to Faulkner, Chick's very youthfulness pointed to the heart of the problem. "It's the grown-ups . . . who keep the prejudice alive," Faulkner told Malcolm Cowley in 1948. "If the race problems were just left to the children," he thought, "they'd be solved soon enough."[10] Chick had already heard as much from old Ephraim, another black friend. Through Ephraim, Faulkner added a second category to the children he had spoken of. "A middle-year man like your paw and uncle, they cant listen," Ephraim had told Chick. People like that were "too busy with facks"; when a person needed "to get anything done outside the common run," he had better try "young folks and womens" because "they aint cluttered"[11] (*ID,* 68, 71).

Aleck Sander and aging Miss Eunice Habersham joined Chick, prompted by familiar motives: Aleck and Chick had been raised like brothers, Miss Habersham and Molly Beauchamp like sisters.[12] Their midnight adventure showed that the bullet that killed Vinson Gowrie had been fired from his brother Crawford's Luger. The effect of this knowledge on the lynch mob was devastating; it destroyed their self-image and sent them scurrying home, unable to face their shame. Those rednecks were "running from themselves," Gavin told Chick (*ID,* 202); but more important, their feelings about Lucas had been turned inside out. Lucas was "now tyrant over the whole county's white conscience" (*ID,* 199), and the mob's guilt, extended to include all blacks, would be an active deterrent to such lynchings in the future. Chick's education on race was now complete. Lucas and Gavin had taught him what produced a mob; he had shown them how to disperse one.

II

With Chick's heroism Faulkner reached a climactic moment in his search for a South. Chick had gone farther than any protagonist Faulkner had yet conceived. He had not only had the courage to break through his inherited prejudices and achieve some understanding of the race problem, but he had also been able, in some measure, to bring about change. Now at last Faulkner saw hope for a South redeemed from its guilty past. Chick, Aleck, and Miss Habersham were prototypes of a new South, a fearless, level-headed, new generation of blacks and whites guided by an older one steeped in tradition. "It doesn't take many," Gavin told Chick;

"three were enough" this time. And Chick, secure in the knowledge that he himself had shown the way, could stand with his people as he had wished: "one unalterable durable impregnable one: one people one heart one land" (*ID*, 244, 210).

Faulkner was fifty when *Intruder in the Dust* appeared, nearing the age when careers wind down; he had reason to hope that this visionary South would stand as a lasting achievement. But talking to a class at the University of Mississippi the previous year he betrayed a strange foreboding. When he was asked who his "most important contemporaries" were, he surprised everyone by listing Thomas Wolfe first, himself second, John Dos Passos third, and Ernest Hemingway fourth. Wolfe had "had much courage"; the North Carolinian had been willing to take chances and "wrote as if he didn't have long to live." Hemingway, on the other hand, had "no courage, has never climbed out on a limb." Rating himself second seemed to mean that Faulkner wanted to be rated the way he rated Wolfe: on the "courage" he had shown taking chances rather than on his success in reaching his goals; but whatever it took to be "important" among his contemporaries, Faulkner did not think he had much of it left. When he was asked what was "the best age for writing," Faulkner responded that "35-45 is the best age for . . . novels"; at that age the "author knows more" than he had earlier, he said, and the "fire" was "not used up" yet. Faulkner seemed worried: "I feel I'm written out," he said with disarming frankness, and he added, "You have only so much steam and if you don't use it up in writing, it'll get off by itself."[13]

The concern about aging that Faulkner had displayed earlier was still with him, and it had begun to sound ominous. Worrying about reaching that "period . . . which most artists seem to reach" in their "early fifties" when "they admit . . . there is no solution to life" and asking to be judged on "courage" rather than on "success," Faulkner sounded like a man whose career had somehow slipped away from him. When he wrote to the American Academy in 1950 to accept the Howells Medal for fiction, he sounded almost preoccupied with the possibility. How, he wanted to know, could people measure a writer's works? "None of mine ever quite suited me," he said; "each time I wrote the last word I would think, if I could just do it over, I would do it better, maybe even right." Faulkner had found ways to keep from worrying. "I would tell myself, maybe I'm too young or too busy to decide"—that anyway, "there was always another one" to write, and he would put off making conclusions. "When I reach fifty," he would tell himself, "I will be able to decide how good or not." Now, nearing fifty-three, Faulkner was ready to tell the American Academy what passing fifty had been like. At first, looking back on his work, he had "decided that it was all pretty good"; but that mood had been all too brief, and what followed sounded horrible. "Then in the same instant I realized that that was the worst of all." He had realized that this "meant only that a little nearer now was the moment . . . when I would put it all away forever that I anguished and sweated over, and it would never trouble me anymore."[14]

Winning the Howells Medal should have afforded Faulkner a moment of triumph; instead, he sounded as though he knew something was wrong, and his pessimistic response spelled out the reasons with a precision no critic could have achieved. Agonizing about how he could "do it better, maybe even right," Faulkner sounded for once as though he knew why, for all the courage he had displayed in *Absalom, Absalom!* and *Go Down, Moses,* he had never grappled successfully with the sources of the guilt that had driven Quentin Compson and Isaac McCaslin; admitting that he had always been "too young or too busy to decide" what was wrong, he sounded as though he knew why his works had continued to founder on the same reefs. No wonder Faulkner thought he might be "written out" at fifty, that he had been afraid of the day when "it would never trouble me anymore." Being "written out" did not mean Faulkner was out of "steam"; *Intruder in the Dust* showed he had plenty left. It meant that during those years that were "the best age for . . . novels" Faulkner had not used that "steam" as he had set out to use it; and now, as he struggled to be "articulate in the national voice," the reasons for his failure were going to be laid cruelly bare.

The sources of that failure were predictable now. Chick was supposed to be guided through the rites of passage by familiar figures: an aristocrat and a black; but the truths about aristocrats and blacks were just the things that Faulkner had never been able to face. That he had chosen Lucas Beauchamp as his model of black manhood revealed much. As a black who had learned how to maintain self-respect, Lucas was right for Chick, and again Faulkner had deftly managed the task of making this paragon of black virtue human; but the very fact that Faulkner had chosen Lucas suggested that he had never engaged the questions that he had not dealt with in *Go Down, Moses.* The problem was still how black Lucas was. The average Yoknapatawpha black could ill afford the dignity of refusing, as Lucas did, to say "mister" to white men or the satisfaction of challenging whites who harrassed him. Lucas was, in fact, far from average. He had never cropped on shares or hired out as labor; he had not only inherited a small farm but (in *Go Down, Moses,* at least) a substantial sum of money; and, as Chick himself acknowledged, Lucas had always had the protection of powerful white cousins.

Lucas never identified himself with the black community. When the old man saw Chick examining a photograph of his wife Molly without her usual head cloth, his comment showed what he thought about blacks. "I didn't want no field nigger picture in the house" (*ID,* 15). When one redneck called Lucas a "biggity . . . Edmonds sonovabitch," Lucas's answer showed where his heart was. "I aint a Edmonds," he said. "I dont belong to these new folks. . . . I'm a McCaslin." When the man threatened him with violence, Lucas answered blandly, "Yes, I heard that idea before. And I notices that the folks that brings it up aint even Edmondses" (*ID,* 19).

This episode effectively dramatized Lucas's bravery and composure; but it did little for Chick's credibility as a southern hero. Chick could learn about white prejudice from a man like Lucas; but learning what being black meant was something else again. Lucas was a familiar compound; his "black" side was a pastiche of Faulkner's notions of the "pride" and "endurance" he thought blacks had learned through "suffering," superimposed on the intelligence and assertiveness that Lucas had inherited from the McCaslins. As before, nothing else was considered, and hence Chick's sense of guilt, on which his heroism was structured, had little to do with the quality of black life. Gavin's observation that Lucas was "a gentleman" identified the source of that guilt; Lucas's concern, Chick realized, was that Chick "had gone out there . . . as the guest not of [Roth] Edmonds but of old Carothers McCaslin's plantation and Lucas knew it when he didn't" (*ID,* 240, 17). *Intruder in the Dust* was not the story of how a black man's heritage could inspire a white youth to rejuvenate his white heritage; it showed, rather, how a white aristocratic youth might be shamed by a part-white aristocratic adult into accepting his white heritage.

That was chilling in its implications for the role of Faulkner's primitives in his vision of a redeemed South; but, when Faulkner turned to the aristocrat who was supposed to be Chick's guide as he applied the inspiration Lucas had taught him, the implications were even more chilling. Some of Faulkner's problems with Gavin emanated from a choice he had made when he conceived the novel's structure. In *Absalom, Absalom!* and *Go Down, Moses* he had drawn on a time-honored device of southern fiction to explore the Old Order's past: a dialogue between two of his characters over the nature of southern experience. Why not employ the same device to explore the South's future? Shreve McCannon's and Cass Edmonds's skepticism had faithfuly revealed Quentin's tragic weakness and Isaac's strength; the prospects seemed good that in similar dialogues Gavin might serve as an effective foil for Chick. But Faulkner was faced with a different problem now. It was one thing to leave Canadian Shreve haranguing conservative Quentin about racial mixture, or to leave conservative Cass brutally but efficiently presiding over the plantation while Isaac retreated into exile; these novels were tragic, and protagonists like Quentin and Isaac achieved pathos because they were unable to reconcile their affection for the Old Order with their fear of its guilty past. But a dialogue reconciling the extremes that Shreve and Quentin, Isaac and Cass, stood for, was another matter. Gavin's function was to bring Chick, crushed with shame at his fellow whites, back into the fold. Chick had to end up agreeing with the opposing voice in the dialogue rather than rejecting it, and that meant that Faulkner had set himself the impossible task of writing a dialogue without a conflict.

Faulkner's solution was in some ways shrewd. He internalized the conflict in Chick, making it stand for the boy's struggle toward awareness. Faulkner could then elevate the dialogue into a joint resolution of Chick's conflict by Gavin and his pupil. But that choice presented more difficulties than it resolved. Chick and Gavin agreed so often, and so completely, that Chick's shame at his heritage seemed, finally, unreal. Faulkner even allowed Chick to notice this situation: often when Gavin would interrupt him, Chick reflected, "without surprise he saw his thinking not be interrupted but merely swap one saddle for another" (*ID*, 153). The lack of disagreement on these issues was devastating for the novel's credibility. Chick was uninterested in finding out the truth about his embarrassing doubts and was interested only in continuing his old loyalties. The effect on Gavin was worse. To make Gavin capable of dominating such a dialogue, Faulkner had to make him a paragon of patrician virtues. Faulkner had taken pains to humanize his knightly lawyer. Gavin was guilty as anybody of assuming Lucas had killed Vinson, and he showed a ready anger at the old man's irascible ways; but as a small-town philosopher who volunteered opinions on everything from racial homogeneity to the urban identity crisis, Gavin was too good to be true, and his virtuosity destroyed what was left of Faulkner's dialogue. It was that virtuosity of Gavin's that finally showed what had gone wrong. Faulkner wanted more from Gavin than a guide for Chick; he wanted a theorist who could prophesy the future of his redeemed South as it took its rightful place in the nation again.

III

The ideas of his characters, Faulkner told Malcolm Cowley in 1948, were not necessarily his own; each had his own voice and spoke to Faulkner from his own point of view. "I listen to the voices," he said, "and when I put down what the voices say, it's right." Faulkner, Cowley reported, had told him that that applied to Gavin, too: "Stevens . . . was not talking for the author, but for the best type of liberal Southerners."[15] Within a decade, forced by the pressure of the impending crisis, Faulkner would announce almost every one of Gavin's and Chick's views as his own;[16] but *Intruder in the Dust* had already made it clear what Faulkner had in mind for Gavin. In 1877 Lamar and his associates had "redeemed" Mississippi from the first Reconstruction; in 1948 Gavin was going to redeem it from the new Reconstruction that southerners were sure Truman's civil rights legislation signaled.

Gavin had taken over Miss Jenny Du Pre's role as spokesman for the Cavalier patriotism of the Lost Cause, and he had no sense of the ironies that a younger Faulkner had seen in Miss Jenny's vision of a "glamorous fatality" in *Sartoris.* That cause was going to remain an inspiration, he insisted. There was a moment "for every Southern boy fourteen years old . . . when it's still not yet two o'clock on that July afternoon in 1863." The afternoon Gavin had in mind was July 3, and the battle was Gettysburg. Gavin pictured "the brigades . . . in position . . . and the furled flags . . . already loosened to break out," and he could see "Pickett himself with . . . his hat in one hand probably and his sword in the other"; Pickett was "looking up the hill waiting for Longstreet to give the word" to advance. At that precarious moment Chick and his people had lost nothing, Gavin said, and they had the world to win. "And that moment," he added, "doesn't need even a fourteen-year-old boy to think *This time. Maybe this time*" (*ID*, 194–95).

Sartoris had been tough about Cavalier chauvinism; as Miss Jenny strolled among the tombs of the Sartoris men, it had been plain that "glamorous fatality" was, in fact, fatal. Faulkner's new Redeemer seemed blissfully unaware of that, and he had little more sense of irony as he moved to the problems that had traumatized Quentin and Isaac. Southerners ought to remember that vision of Pickett's charge, Gavin said, because it validated their sense of ethnic solidarity.[17] Gavin had a more sophisticated name for the racial "purity" that Isaac had looked back to: "homogeneity." There are "a few of us," he told Chick, who "know that only from homogeneity comes anything of a people . . . of durable and lasting value." He was talking about everything viable in a culture: "the literature, the art, the science, that minimum of government and police which is the meaning of freedom and liberty," and, most important, "a national character worth anything in a crisis." After all that, it came as no surprise where Gavin thought such homogeneous people could be found. Southerners "alone in the United States," he announced, "are a homogeneous people" (*ID*, 154, 153).

Gavin was trying to be objective. He did not just mean white southerners, he told Chick; the black, whom he disconcertingly referred to as "Sambo," had his homogeneity, too. In Clytie Sutpen, Rosa Coldfield had sensed "an older and a purer race than mine"; Roth Edmonds had thought Lucas Beauchamp was "a man most of whose blood was pure ten thousand years when my own anonymous beginnings became mixed enough to produce me." Gavin was ready to admit that "Sambo" had inherited "a better homogeneity than we have," and he thought that had produced specific virtues. These resembled the ones Isaac had identified. The black had "had patience even when he didn't have hope," Gavin told Chick, and he had shown "the desire to endure"; he had been able to take "the long view even when there was nothing to see at the end of it." Blacks loved "a God," Gavin thought, and, if he did not seem to be the same God whites worshiped, he offered a spirituality at least as valuable: "a heaven which a man may avail himself a little of . . . without having to wait to die" (*ID*, 155, 156).

Gavin's notions of black virtue revealed the problem Faulkner was facing with his Redeemer. Isaac, announcing similar virtues, had been dealing with the world of 1888, and, if he seemed at times to be describing no more than the qualities of a reliable slave, his romantic affirmation of black sensibility was a challenge of his own white ethic. Gavin was speaking to the realities of another century, and he sounded regressive. For Isaac, moreover, those virtues had been the end-product of a primitivism that travestied white civilization; Gavin's remark about a heaven on earth was his only hint about primitivism. He never suggested that the black virtues he identified were more than the result of "homogeneity," and that made all the difference. It made blacks appear circumscribed by those virtues rather than liberated by them.

There were ironies for Faulkner that ran deeper than those Gavin represented; allowing his Redeemer to base his black virtues on "homogeneity" was a strange twist. In the world of Faulkner's youth, apostles of racial purity like James Kimble Vardaman and Thomas Dixon, Jr., had hounded blacks as the enemies of white racial purity, ringing down an iron curtain of segregation to prevent an "amalgamation" that signaled "the end of this nation's life"; now Faulkner had Gavin telling blacks, in effect, that they had to keep their own racial purity to preserve their virtues, and it sounded like another form of segregation. Not all the ironies of that situation emanated from the past, either. It was clear how valuable homogeneity was, Gavin told Chick, from what happened where people did not have it, and to show this he pointed to the regions that were providing the impetus for Truman's civil rights legislation.

Faulkner used Chick's youthful impressions about the North to soften this attack. Chick "had fed from his mother's milk" on an attitude toward the North; it had become not so much "a geographical place" for him "as an emotional idea." He had been trained to think of the North as "outland and circumscribing," something he had "to be ever and constant on the alert" for. His training told him he was "not at all to fear" the North "and not actually anymore to hate" it; he just had to be ready "to defy" the North—even if he did it "a little warily sometimes and sometimes even with tongue in cheek." His training was at the root of a comic fantasy that had been with Chick since "infancy": "a childhood's picture" of "a curving semicircular wall . . . from the top of which . . . there looked down upon him and his countless row on row of faces." Those "resembled his face and spoke the same language he spoke," but between them "and him and his there was no longer any real kinship"; even "the very mutual words they used would no longer have the same significance." Before long, there would be "only the massed uncountable faces looking down at him and his in fading amazement"—an amazement symptomized by an "almost helpless . . . eagerness to believe anything about the South not even provided it be derogatory but merely bizarre enough" (*ID*, 152–53).

Gavin wanted Chick to know such fantasies had their basis in reality. The problem was that northern cities had been taken over by "the coastal spew of Europe" and lacked the "national character" homogeneity brought. Such outlanders "no longer . . . [had] anything in common save a frantic greed and a basic fear of a failure of national character"; because of this the North had "had to surrender voluntarily more and more of its . . . liberty in order . . . to afford the United States." Heretofore, these people had been "quarantined . . . into the rootless ephemeral cities with factory and foundry and municipal paychecks"; but conditions were changing now, and what was happening sounded disconcertingly familiar. In 1866 a hostile Congress had passed the first Reconstruction; now these "outlanders" were threatening a new Reconstruction in the name of civil rights. They were "forcing on us laws" based on a misconception: "that man's injustice to man can be abolished overnight by police" (*ID*, 153, 155, 156, 203–4).

Gavin's preachments showed how far the integration crisis had carried Faulkner; there was a desperation about those words, as though, if Faulkner did not get everything explained properly, a terrible tragedy would follow. Northerners had forgotten the lessons of the Civil War, Gavin said. What had happened to Lucas was wrong; but "outlanders" ought not to forget that there had been a time when "Lucas Beauchamp's freedom was made an article in our constitution and Lucas Beauchamp's master was . . . trampled for ten years on his face in the dust to make him swallow it." If outsiders could make the South's redneck elements accept black equality, they would have been able to do it then; trying it now would only arouse the rednecks against blacks again and "fling . . . [the black] decades back" in his efforts to free himself. It would fling him "back not merely into injustice but into grief and agony and violence too" (*ID*, 155, 203).

Outlanders did not understand the South. There might not be one individual "in any random one thousand Southerners," Gavin insisted, who was "really concerned" about injustice to blacks, and not "always one who would himself lynch Lucas." But one thing was certain: outside intervention would unite the entire one thousand, and the resulting North-South confrontation would be very different from what northern liberals thought. There would be "a concorded South" on the one hand, and Gavin had harsh words for those on the other: they would consist of "a paper alliance of theorists and fanatics and private and personal avengers," united with "others under the assumption of enough . . . [distance] to afford a principle" (*ID*, 215–16). Still worse, many "ignorant" northerners "who fear the color of any skin . . . save their own" would join the southern side. The southern "liberal" like Gavin, his position undermined, would be forced into alliance with the worst type of reactionary; the Civil War, in short, would be refought, and there would be no winners.

That, Gavin wanted it known, was where northern liberals were mistaken about southerners. They thought the South was just aligning itself against "progress and enlightenment"; but southerners saw themselves in other terms. It was "simply our homogeneity" they were defending from the federal government, Gavin said, not "our political beliefs or even our way of life," and there was a lesson there for southerners. Black homogeneity had to be preserved, too; blacks' "capacity to survive and absorb and endure and still be steadfast" had to be protected. Southern blacks and whites had more in common than either group had with northerners; those two great homogeneous groups ought to "confederate." Whites ought to "swap . . . [the black] the rest of the . . . privileges which are his right, for the reversion of his capacity to . . . endure and survive." That, certainly, was a new kind of Confederacy. Gavin's sermonizing had taken him full cycle. Defending the South, he had said, meant defending blacks from southerners; the South had to clean its house if it was going to be taken seriously. But in the larger context of "homogeneity," defending blacks meant defending the South. "I'm defending Sambo," he concluded, "from the North and East and West" (*ID*, 153, 204, 156, 203).

IV

Gavin's final exercise in doublethink showed what had happened to

Faulkner's visionary South. If Faulkner was listening to the "voices" of his characters now, as he had told Malcolm Cowley he was in 1948, those voices were not telling it straight; what he was hearing was the voice of the world of his youth, and, as it struggled with the problems of a new era, that voice sounded strained, almost hysterical. Far from helping Chick separate the virtues of his patrician ethic from its Cavalier chauvinism, Gavin seemed to be making that impossible. Chick and Gavin seemed uneasy when they were merely identifying themselves with their "homogeneous" southern contemporaries; they wanted to identify against outsiders. That revealed itself as Chick recalled what Gavin had told him about Pickett's charge. This chauvinistic daydream did not suggest any liberating victory; its goal was dominance. Chick was thinking about what southerners had had "to gain" in that battle, and his mood was ambitious: "Pennsylvania, Maryland, the world, the golden dome of Washington itself." Even Gavin's daring suggestion that southern whites and blacks ought to "confederate" to defend their homogeneity smacked of militarism. So allied, he argued, they would not merely "present a front . . . not even to be threatened"; they "would dominate the United States" (*ID,* 195, 156).

In *Light in August* Gail Hightower's fantasies of military glamour had been counterpoised against the tale of his Cavalier grandfather's ignominious death in a chicken coop and Percy Grimm's demonized Americanism balanced against Joe Christmas's lonely martyrdom; but Gavin's and Chick's belligerence had nothing to balance it. Often it seemed downright paranoid. Gavin imagined himself debating a suitably prejudiced northern liberal who could not accept his belief that southern blacks wanted to avoid a confrontation. "You must know Sambo well," this opponent sneered, "to arrogate to yourself such calm assumption of his passivity." That allowed Gavin to reveal his own sophistication. "I don't know him at all," he insisted, impaling his opponent on one of the oldest of racial clichés, "and in my opinion no white man does." When Gavin argued that blacks had to wait until the race issue cooled, his antagonist insisted outsiders had to give them freedom now "because you not only cant you wont." That allowed Gavin to cast himself as the voice of reason. "Come down here and look at us," he suggested, "before you make up your mind." The reply, he fancied, would be hostile: "No thanks the smell is bad enough from here." Gavin's devil-liberal, supreme in his self-righteousness, had closed his mind; he was not only ready for war, he was ready to congratulate himself on his motives. "At least we perish in the name of humanity," he proclaimed, allowing a level-headed Gavin a last ironic word: "what price Lucas' humanity?" (*ID,* 215–17). With that paranoid fantasy, Gavin's mask was, finally, off.

Faulkner's Redeemer was a repository of prickly, threadbare hostilities. They undermined whatever effectiveness he might have had as Faulkner's spokesman and, with it, what was left of Faulkner's vision of a redeemed patrimony. Gavin's belligerent fantasies showed why passing fifty had troubled Faulkner so deeply; they revealed, as nothing else could, the wrong road that he had taken in his search for a South during those years he had thought were the "best age for writing novels." It was too late now for Faulkner to retrace his steps; that road had taken him into a shadowy region where he could not find his way, and now, as he struggled to reach the visionary South that was to provide the climax of his career, he found that it was beyond his grasp. Sadly, Faulkner seemed dimly aware of what had happened. The clue, perhaps, was a strange indifference that seemed to pervade Faulkner's visionary novel; it was as though, poised to show how his patrimony might be redeemed, he felt almost that it was not worth the trouble.

Faulkner had never been a master at tying up the loose ends of his plots; but there were so many of those in *Intruder in the Dust*, it was almost beyond belief. Gavin was supposed to be an aristocrat, and it was true that aristocrats in the world of Faulkner's youth could be as concerned with "homogeneity" or its equivalents as anyone; but it had been redneck leaders like Vardaman and Dixon who made racial purity their platform against both aristocrats and blacks. Such speeches sounded out of place coming from Faulkner's aristocrat. What was more, Faulkner ought to have been embarrassed by Gavin's inconsistencies on the subject. Roth Edmonds knew the truth about "homogeneity," that his "own anonymous beginnings" were fatally "mixed"; and Isaac McCaslin, unable to face the change interbreeding meant, was still aware that in the Delta "Chinese and African and Aryan and Jew, all breed and spawn together until no man has time to say which one is which nor cares." Faulkner's Redeemer seemed unaware of that; and he was on ground just as shaky claiming "a better homogeneity than we have" for American blacks. Gavin's knowledge of Lucas ought to have suggested to him that American blacks usually have some white ancestry, even if he was not aware (as Isaac and Roth also did not seem to be) that Africa was a diverse continent, nurturing many ethnic strains.

There were deeper problems, too. Faulkner began to display surprising weaknesses where he ought to have been strongest.[18] A murder mystery was at best a doubtful vehicle for his vision of a redeemed South; but, once he had decided to write about "a nigger in his cell, trying to solve his crime," he could not find it in him to pay enough attention to the conventions of the mystery to make his novel credible. Faulkner, who had been credited with the screen play for Howard Hawks's *The Big Sleep*, knew mystery writers characterized their murderers in at least minimal detail in the beginning of a novel or movie, so that when their identities were revealed those murderers were believable criminals. But Faulkner never showed Crawford Gowrie in detail, and it came as an unwelcome shock that this simple Beat Four rowdy was capable of killing his brother over a few loads of lumber. Worse, Faulkner relegated that plot to the background as Chick's heroism and Gavin's theorizing emerged. Finally, instead of dramatizing the mystery's resolution, he allowed Gavin to narrate it as an afterthought. With his mystery's motivation and solution both virtually thrown away, Faulkner's plot was revealed for what it was: a thin subterfuge that allowed Gavin to talk. It was as though, his eyes on the novel's polemics, Faulkner could not bring himself to come to grips with its dramatic consequences.

This indifference to the implications of his plot implied a great deal about Faulkner's future. It was as though he had begun to despair of fiction as the vehicle for his search for a South, as though he were looking for some other way to continue that search. The next decade of his life would confirm these changes. As the crisis over integration became critical at home and the Cold War became a cause for concern abroad, Faulkner would launch himself on a new phase of his search: a round of speechmaking designed to set before the world the ideas that Gavin had conveyed to Chick. But even that crusade would not explain the indifference Faulkner had shown in *Intruder in the Dust*; it would only suggest its motivation.

The explanation was closer to home. In *Absalom, Absalom!* and *Go Down, Moses,* Faulkner had covered himself subtly as he evaded Clytie's secret and Carothers McCaslin's immorality, manipulating his plots by making Thomas Sutpen and Old Carothers sound like rednecks instead of planters. In *Intruder in the Dust* he was still manipulating his plots to justify Gavin's polemics, but his manipulations bordered on the ridiculous, and they were there for all to see. The most hopeful implication of *Intruder in the Dust* was Faulkner's vision of how brave, concerned people could stop a lynching. Faulkner had come up with a suitably heroic episode to illustrate his point: Chick's, Aleck's, and Miss Habersham's midnight excursion into the Beat Four graveyard. As Gothic adventure, the episode was compelling; the problem was its value for Faulkner's vision of a redeemed South. First, the episode was unlikely; sixteen-year-old boys and old women seldom go around digging up graves, much less those of testy hill clans. For another, it was too easy; compared to Henry Hawkshaw's and Gail Hightower's futility with similar mobs in "Dry September" and *Light in August,* the success of these unlikely heroes seemed miraculous. In *Go Down, Moses,* stereotyped episodes like Lucas's gold hunt or Gavin's sentimental funeral for

Butch Beauchamp had a direct bearing on Isaac's repudiation of his heritage; but one only had to imagine a youthful Isaac digging in a midnight graveyard to save Sam Fathers, or Cass Edmonds explaining a murder, to realize what had happened. If this was the most hopeful way Faulkner could imagine to forestall lynchings, the South's problems seemed far from over.

That flimsy resolution of Lucas's troubles was a strange development; it was almost as though Faulkner were trying to beg the very question he claimed to be resolving. Other loose ends were more damaging. If Chick's heroism could defuse Lucas's lynch mob, Gavin wanted it known, any mob could be handled; but here Faulkner seemed to have lost all sense of irony. Will Mayes in "Dry September" was lynched for a rape that had never happened, and the irony of Christmas's death was that he had killed a white woman and wanted to be lynched. But Faulkner had forgotten the complexities of those earlier tales, and he was willing to introduce a side issue that begged the question even more: Vinson Gowrie's murderer, it turned out, was his own white brother. Thus, the shame that drove the mob out of the Jefferson streets was not shame that they had nearly killed the wrong man, nor was it shame that a white man, rather than Lucas, had committed that crime; it was shame that a white man had committed fratricide. That fact fatally altered the meaning of Chick's heroism; it implied Lucas could be saved only if a crime such as fratricide was involved.

It was appalling that Faulkner had introduced a side issue of such complexity at this moment; it was as though he felt compelled, now that he seemed on the point of reaching the visionary South he had struggled for, to show that he had not found it after all. But Faulkner was not through. Toward the end of the novel he contrived to beg the question of Chick's heroism even further when he had Chick suddenly remembering that it had taken the Gowries three days after Vinson's killing to come to town for Lucas; if they had wanted Lucas, Chick realized, they could have had him at almost any time. For Chick, that fact threw Lucas's troubles into a new perspective. "The Gowries themselves had known he hadn't done it," he realized; aware that the old man knew the murderer's identity and they did not, they were just "waiting for somebody else . . . to drag him out into the street until he remembered" (*ID*, 219).

If pursued, that incredible revelation would undermine almost every other implication of the book. For all Chick's heroism, Lucas had never been in danger of being lynched by the Gowries; and since, as Gavin insisted, no other element in the mob was capable of a lynching, there never could have been any! Chick's revelation, coming in the novel's closing chapter, was the final incredibility in an incredible performance.

The writer who had shown Will Mayes's and Joe Christmas's lynchings as the inevitable results of poor-white scapegoating rituals had manipulated Lucas's troubles to prove that such rituals were not dangerous. In doing that, he had also shown that perhaps there was no need to write a novel about Chick's heroism.

Those strange ambiguities disclosed much about Faulkner's fears as he reached fifty. *Intruder in the Dust* cried out to be judged as Faulkner had judged Wolfe, on the "courage" it took to take "chances" rather than on whether he had reached the goals he had set for himself. It was as though, by undermining his own plot, Faulkner was trying to tell people something he could not admit even to himself: that during those years that were "the best age for writing novels" he had not paid the price of his climactic vision of a redeemed South, and that, because he had not, his vision was built on quicksand. Chick's heroism seemed to speak for the Faulkner who had begun *Absalom, Absalom!* and *Go Down, Moses* in bouts of determined introspection; but those ambiguities spoke for another Faulkner, a Faulkner who felt none of his novels "ever quite suited me," who confessed that "each time I wrote the last word I would think, if I could just do it over, I would do it better, maybe even right." That was the Faulkner who at fifty had thought he was "written out," who feared a "period in their early fifties which most artists seem to reach" when "they admit at last that there is no solution to life and that it is not, and perhaps never was, worth the living." That Faulkner was going to be a liability as he moved away from fiction to crusade for the values for which Gavin and Chick had stood.

11

Doing Something about It

I

In May 1951 Faulkner addressed his daughter Jill's high school graduating class. "Our danger," he said, "is the forces in the world today which are trying to use man's fear to rob him of his individuality, his soul." The class of 1951, he said, could do something about that. "Never be afraid to raise your voice for honesty and truth and compassion," he told them, and "against injustice and lying and greed." He was ready to issue a challenge. If the seniors in that room would join those "in all the thousands of other rooms like this one . . . today" and stand up "as individuals, you will change the earth."[1]

Faulkner was about to embark upon the most remarkable episode of his search for a South: he was ready at last to step from behind the facade of his fiction and take over Gavin Stevens's role as Redeemer. His Nobel Prize acceptance speech in 1950 had sounded a note of hope in a time of anxiety: "I decline to accept the end of man. . . . I believe that man will not merely endure: he will prevail. He is immortal . . . because he has a soul, a spirit capable of compassion and sacrifice and endurance."[2] To the U.S. State Department, the address was an unexpected windfall. Soon Faulkner was circling the globe as goodwill ambassador, struggling to tell uncommitted nations about America; and, unsatisfied merely "to raise . . . [his] voice for honesty and truth and compassion" abroad, he was also ready to raise it "against injustice and lying and greed" at home. He seemed determined to go public with every aspect of his beliefs and just as determined to let no crisis pass without involving himself.

Later, as Faulkner answered an inquiry about the heroes of his novels, he seemed to be explaining what had moved him in those years. There were three "stages" in the evolution of people who knew how to "cope with the problems," he said. "The first says, This is rotten, I'll have no part of it, I will take death first." That sounded like Quentin Compson, and Faulkner identified the second as Isaac McCaslin: "The second says,

This is rotten, I don't like it, I can't do anything about it, but at least I will not participate in it myself." The final stage sounded like Chick Mallison in *Intruder in the Dust*. "The third says, This stinks and I'm going to do something about it."[3] Faulkner had spent a long career identifying what "stinks," trying to find heroes who could "do something about it"; now, at fifty-four, he must have felt that he was in an extraordinary position to achieve both of those things himself. As a southern aristocrat, he was at home with the traditional values Gavin Stevens thought "outlanders North East and West" had lost. And as a novelist, he could also tell himself that he had achieved what few southerners had attempted: in Chick and Gavin he had created heroes who could reaffirm those values. Now the final challenge of his search for a South loomed suddenly before him, and he was ready to respond with the same tragic commitment with which he had responded in his fiction.

Back in Memphis in 1955 after a tour for the State Department, Faulkner told the Southern Historical Association what he thought Americans were facing. Of the nine countries he had visited, "the only one" he was sure would "not be communist ten years from now, is England." He was sure that "if all the rest of the world becomes communist, it will be the end of America too as we know it." People had to realize that "the only reason all those countries are not communist already, is America." But America had to clean its house in order to justify its leadership. "If we who so far are still free, want to continue to be free, all of us . . . had better . . . confederate fast, with all the others who still have a choice to be free." To Faulkner, the key battle of the Cold War was going to be fought in the South; many of the uncommitted nations were black, and, if America was going to appeal to them, it had to do something about the way blacks were treated at home. If southerners did not come to grips with this problem, Faulkner warned, the South might "wreck and ruin itself twice in less than a hundred years, over the Negro question." Southerners had to realize that their future reduced itself to living with racial equality: "To live anywhere in the world of A.D. 1955 and be against equality because of race or color, is like living in Alaska and being against snow."[4]

That address identified Faulkner's problem. The Faulkner who called for national and regional housecleaning was familiar from his fiction; that Faulkner had begun *Absalom, Absalom!* and *Go Down, Moses* ready to probe the abolitionist as well as the southern beliefs about the southern past. But there was another Faulkner, the one who felt none of his novels "ever quite suited me," who confessed that "each time I wrote the last word I would think, if I could just do it over, I would do it better, maybe even right."

The ideas of his characters, Faulkner insisted, were not necessarily his own. He was not sure "any writer could say just how much he identifies himself with his characters," he said in 1957 when somebody asked him about Gavin Stevens. Yes, there were "moments when the character is in a position to express truthfully things which you yourself believe to be true," but that did not mean "that you're trying to preach through the character"—it was merely an accident. "It just happens that this man agrees with you on this particular point."[5] But, as Faulkner threw himself into his new role, Gavin, Chick Mallison, and Isaac McCaslin agreed with him more often than not, and sometimes Cass Edmonds agreed with him, too.

Faulkner's 1955 efforts for the State Department in Japan revealed the ambiguities of his commitment. Under the pressure of the Cold War, he tried to downplay his regional loyalties, and that was hard. Yes, he admitted, America had not always been united. "One hundred years ago, my country, the United States, was not one economy and culture, but two of them." In the resulting war, "my side, the South, lost." The bitterness on his "side" was real: "we were not only devastated by the battles which we lost, [but] the conquerer spent the next ten years after our . . . surrender despoiling us of what little war had left"; the northern states "made no effort to rehabilitate and reestablish us in any community of men or of nations." What Faulkner wanted the Japanese to understand was that "all this is past: our country is one now."[6]

But the more Faulkner talked, the more anxious he seemed to be to explain his southern loyalties to the Japanese. More often than not, when he spoke to them about "my country," he meant the same thing he had meant when he said "my side." That was not just the South, but the deep South: "most of my country lies between New Orleans and Memphis."[7] The ghost of the parliamentary wars seemed to hover over that country. "My people were Scottish," he said. "They fought on the wrong side [in Scotland], and they came into America, into Carolina, and my grandfather"—he meant his great-grandfather—"chose the wrong side again in 1861."[8] As Faulkner explained why his southern countrymen had a strong sense of family, he sounded like Gavin explaining the redneck geopolitics of Beat Four. "We have to be clannish," he insisted, "just like the people in the Scottish highlands, each springing to defend his own blood whether it be right or wrong." The frontier and the war had done that, he said. As he went on to elaborate where Gavin had only hypothesized, Faulkner sounded like he was working up a rationale for Gavin's notion of "homogeneity." Clannishness was "regional"; it could be found throughout the South among "industrialists" as well as "farmers."[9] Describing that clannishness forced Faulkner to try to cope with its ambiguities. He felt "hate" as well as "love" for his "country," and his affections had been divided for some time. Faulkner wanted the Japanese to know that he was aware of what the South's problems meant to non-whites. "There are things in my country that I don't like," he admitted. If you loved "a land, a people, or even a person," it was "not because of what that person or land or place is"; you loved it "despite" those things, and he wanted the Japanese to know he had devoted his life to doing something about them. "The only way that I can cure its faults within my capacity, within my own vocation, is to shame it." He had to make people "angry enough, or shamed enough to change it."[10]

Through the mid-1950s Faulkner's desire to "cure" those "faults" took the form of an almost compulsive desire to publicize his ideas about race. Theorizing about black and white ethnocentricity to London *Times* reporter Russell Warren Howe in 1956,[11] he insisted that all races had equal potential. "There is no such thing as an 'Anglo-Saxon' heritage and an African heritage," he said; there was only "the heritage of man." He was sure that "nothing . . . [was] extinct in any race," it was "only dormant." But as Faulkner told Howe about the values of a black American heritage, it became plain that blacks, at least, had characteristics that seemed "dormant" in whites. Like Isaac, who thought black vices were those "white men and bondage have taught them," Faulkner insisted that "the vices that the Negro has have been created in him by the white man"; but their virtues seemed to be their own homogeneous ones and resembled some of those Isaac had identified. Isaac had thought blacks were "better than we are" and "stronger than we are"; Faulkner wanted Howe to know blacks were "calmer, wiser, more stable than the white man."[12] In Japan he came down hard on Isaac's notion of endurance. The black "won't vanish as the Indian did," he said, because he had "a force, a power of his own that will enable him to survive." The black's condition in America had given him that: "The Negro is trained to do more than a white man can with the same limitations." And his endurance had given him stature. "To have put up with this situation so long with so little violence," he told Howe, "shows a sort of greatness."[13] On three occasions he invoked a kind of apostrophe to what he saw as black progress. In "only three hundred years in America," he pointed out, blacks had "produced Ralph Bunche and George Washington Carver and Booker T. Washington." They had "yet to produce a Fuchs or Rosenberg or Gold or Greenglass or Burgess or McClean or Hiss, and for every prominent communist or fellow-traveler like Robeson," he insisted, "there are a thousand white ones."[14]

Faulkner was trying to be tough about the justice of black aspirations. The condition of blacks was deplorable. There was a "simple incontrovertible immorality" about "discrimination by race."[15] There could be "no opinion" about what the Supreme Court had done, "it's right, it's just."[16] He was for nonviolent pressure on segregated institutions, he told the readers of *Ebony* in 1956, and, although he felt that the National Association for the Advancement of Colored People ought to avoid pushing into hopeless situations, he had high praise for it; if he were black, he would join, "since nothing else in our U.S. culture has yet held out to . . . [blacks] so much of hope." But what he favored was more like what a young minister named Martin Luther King, Jr., had been trying in a bus boycott in Montgomery, Alabama. Blacks ought to pursue "a course of inflexible and unviolent flexibility directed against not just the schools but against all the public institutions from which . . . [they] are interdict." Such a program might "send every day to the white school to which he was entitled . . . to go, a [black] student . . . courteous, without threat or violence." And every time such a student was denied, others would come singly but steadily until the barriers crumbled. Such "was Gandhi's way."[17]

Faulkner responded deterministically to questions about the causes of prejudice. "The racial problem in my country is an economic one," he told the Japanese; it was something that happened when white children "get old and inherit that Southern economy which depends on a system of peonage."[18] Faulkner seemed to have forgotten the curse of Cain and the Manichaean projections of his Puritan poor whites in *Light in August.* Some southerners, he complained, would have people believe that "the Bible says that the Negro must not be our equal, that if the Lord wanted him to be our equal, the Lord would have made him white." Those were "foolish, silly things," Faulkner thought; but now, trying to explain them as economic rationalizations rather than as projected guilt, he seemed baffled and could account for them only as "baseness." He seemed to have forgotten other things, too. People were just as foolish and silly, he said, when "they would tell you that a different kind of blood runs in the Negro's veins." The war of inimical bloods that had ravaged Clytie Sutpen, the benign fusing of bloods that made Sam Fathers and Lucas Beauchamp superior individuals—those things seemed ridiculously simple now. "Everybody knows that blood's blood," Faulkner insisted. He could understand why whites wanted to maintain their status; but why they would create such myths escaped him. "What I don't like is the fact that these people . . . use such base means."[19]

As Faulkner developed his notion of economic bias, it began to sound like class bias. There was "more prejudice in the South" than in the North, he felt, because the South's "economy is less complex." Sometimes he seemed to blame economic prejudice on aristocrats. It was "the bankers who depend on the mortgages on the cotton," he told the Japanese, and "the planters who have got to make the crop, with . . . Negro labor." The southern white man was "afraid that if he gives the Negro any . . . social advancement at all, the Negro man will stop working for the low wage and then he will get less when he sells his cotton." At other times he seemed to be thinking about poor whites instead of patricians. "There's a class of white man," he said, "that hates the Negro simply because he's afraid that the Negro will beat him at his own job, his own economic level." People were afraid "that the black man, if he has any social advancement," would "take the white man's economy away from him." Then "the white man will have to work for less money because the Negro will own the cotton land and the cotton."[20]

In these painstaking analyses of black virtue and white prejudice, Faulkner had set out to identify what "stinks" in his native land. What remained was to find ways "to do something about it." But for the man who had manipulated his plots to avoid forcing Isaac and Chick to face the full impact of the South's guilty heritage, that was going to be difficult. Faulkner, at last, was faced with one plot he could not manipulate.

II

In 1950 Faulkner wrote to the Memphis *Commercial Appeal* about Leon Turner, a white man found guilty by an Attala County, Mississippi, jury of killing three black children.[21] "All native Mississippians," he asserted, "will join in commending Attala County"; but the verdict had a special significance for certain Mississippians, for "those of us who were born in Mississippi and . . . who have continued to live in it for forty and fifty and sixty years at some cost and sacrifice." Loving their state, they had "been ready and willing at all times to defend our ways and habits and customs from attack," and the conviction in Attala made such people feel gratified that they had chosen to remain in Mississippi.[22]

The Turner letter foreshadowed the difficulties Faulkner was facing. There was a vast gap between his expressed belief in a housecleaning that would save America and the South from the Cold War and his fear that people like himself would have to leave the South. Four years later, when the U.S. Supreme Court rendered its decision in *Brown* vs. *The Board of Education* and set off the long-awaited storm, Faulkner set out to persuade as many whites as possible to accept black equality before the North forced it on them. All southerners, he wrote the *Commercial Appeal* in 1955, had to face "two apparently irreconcilable facts." One was "that the National Government has decreed absolute equality in education among all races"; the other was that there were "people in the South who say that it shall never happen." He had a plan. Emphasizing that "this only solves integration: not the impasse of the emotional conflict over it," he proceeded to elaborate the unthinkable: Mississippians ought to look at the values of an integrated school system rather than at the drawbacks. The state's "present . . . reservoir of education is not of high enough quality to assuage the thirst of even our white young men and women," he asserted. "If we are to have two school systems, let the second one be for pupils ineligible not because of color but because they either can't or won't do the work of the first one."[23]

The reaction to his letters was predictable. Most white Mississippians who responded were outraged, and few seemed capable of listening to what Faulkner was trying to tell them. Faulkner patiently answered those responses; but, five months later, nearing the end of a round-the-world State Department tour, he had begun to sound angry himself. Emmett Till, a fourteen-year-old Chicago black visiting in the Delta, had disappeared, apparently killed for making overtures to a white woman. Faulkner, in Rome when he learned of the incident, seemed to have his mind more on those uncommitted nations he had spoken of than on being accepted by Mississippians. When, he wanted to know, were Mississippians going to learn about the outside world and about the Cold War? "When will we learn that if one county in Mississippi is to survive it will be because all Mississippi survives? That if the state of Mississippi survives, it will be because all America survives?" Maybe that killing might "prove to us whether or not we deserve to survive"; but he was sure that "if we in America have reached that point in our desperate culture when we must murder children . . . we don't deserve to survive, and probably won't."[24]

Faulkner's statement was tougher than anything he had yet said; but he had reason to think that its toughness would serve more than one purpose. It might, for example, solidify his position with northern liberals, and he was going to need their approval soon. If his campaign against lynching and segregation had qualified him as a southerner to whom liberals might listen, he would be in a position to help prevent liberals and segregationists from coming to blows. In February 1956 black coed Autherine Lucy attempted to enter the University of Alabama. Riots erupted in Tuscaloosa, and Lucy barely escaped with her life as angry whites stoned her car. To Faulkner, the incident indicated that it was time to direct his appeals to northerners. "I . . . have been on record," he wrote in *Life*, "as opposing the forces in my native country which would keep the condition out of which . . . [segregation] has grown"; now he wanted to "go on record as opposing the forces outside the South which would use . . . compulsion to eradicate that evil over night." Faulkner wanted to caution liberals: "Go slow now. Stop for a time, a moment."[25]

More was at stake than Lucy, it appeared. Five years earlier he had been cautioning Mississippians about driving off people like himself "who have continued to live . . . [in the state] at some cost and sacrifice"; now he wanted northern liberals to realize what was happening to such people. There were many southerners "who believe as I do and have taken the same stand" against discrimination. They were moderates. They were helping their region "by still being Southerners, yet not being a part of the . . . majority Southern point of view; . . . by being in the middle, being in position to say to any incipient irrevocability: 'Wait, wait now, stop and consider first.' " Northern liberals were so bent on forcing integration that they were making life impossible for southern moderates, and Faulkner was thinking about the polarization of southern opinion Gavin had feared. If moderates were "compelled by the simple threat of being trampled if we dont get out of the way," it could be fatal; then "the line of demarcation" might "become one of race," and "the white minority like myself" might be "compelled to join the white segregation majority no matter how much we oppose the principle of inequality."[26]

Faulkner was getting a familiar feeling—the one Chick had when he fantasized "countless row on row of faces" which looked at "him and his" across a chasm that threatened to cleave the two groups "too far asunder even to hear one another." Faulkner wanted northerners to understand the problem. "The rest of the United States knows next to nothing about the South," he insisted. He recalled a New York *Times* editorial that said the Tuscaloosa riots over Lucy's admission were "the first time that force and violence have become a part of the question"; for southerners, that was a naive statement. "To all Southerners . . . the first implication . . . of force and violence was the Supreme Court decision itself." Faulkner believed that the North had not yet learned the real lesson of the Civil War: "that the South will go to any length . . . before it will accept alteration of its racial condition by mere force of law or economic threat." Liberals had to "stop now for a moment. You have shown the Southerner what you can do . . . ; give him a space in which to get his breath and assimilate that knowledge."[27]

III

"A Letter to the North" identified the personal tragedy that Faulkner faced as he struggled with his role as Redeemer. "Being in the middle" was no place for Faulkner. He found himself now exactly where his class had taken its stand since the Redemption: in that twilight world between the twin catastrophes of the Reconstruction and the revolt of the rednecks, where aristocrats struggled to defend blacks and fellow aristocrats against poor whites and to defend all three classes against northern intervention. In his fiction the mature Faulkner, who wanted "to raise" his "voice . . . against injustice and lying and greed," had struggled through a long career to vacate that ground, to understand the guilty past that had brought Puritan rednecks and Puritan Yankees to attack his patrimony; but whenever he had seemed on the point of facing those accusations, his commitment to the world of his youth had interposed, and he had found himself manipulating his characters and plots to avoid them.

Now there was no way to avoid that conflict. The southern response to his integration stand was quick and vindictive; it seemed indifferent to the spirit of Faulkner's efforts and out of proportion to their significance. Letters to the *Commercial Appeal* impaled him on the dulled points of familiar clichés; people attacked his courage, his intelligence, even his education. "When Weeping Willie Faulkner splashes his tears about the inadequacy of Mississippi schools," one wrote, "we question his gumption."[28] Another volunteered that since Mississippians were "only mortal men and not Nobel Prize winners," they had to "bow to . . . [Faulkner's] higher intellect"—then ignored his self-proclaimed unimportance long enough to ask acidly "how many degrees . . . [Faulkner] holds from those inferior Mississippi schools of which he writes."[29] (Faulkner was, he said elsewhere, "an old veteran sixth grader."[30]) John Faulkner recalled that his brother "became subject to anonymous phone calls at odd hours. Mysterious voices cursed him, and his mail was filled with abusive anonymous letters."[31]

Some of those angry people were neither anonymous nor mysterious. "I doubt," Jack Falkner wrote, "that Mother could ever bring herself to believe Bill actually meant what he publicly said." Jack could hardly believe it, either: "I could not understand how he, whose life had been so much like my own, could have arrived at the conclusion he had expressed."[32] John Faulkner was furious; years later, as he tried to recall the family's reactions to those insulting letters and calls, he insisted they had all felt that "it serves him right. He ought to have known this would happen."[33] With his integration stand, Delta liberal Hodding Carter wrote in 1957, "Faulkner ventured beyond the ultimate Southern pale"; Carter thought that "nothing, not even the initial Oxford reaction to *Sanctuary*, could match what has happened to him since." Faulkner's status was reduced to that of "a renegade in his homeland," Carter lamented; southerners saw him only as "Weeping Willie Faulkner, nigger lover, . . . a would-be destroyer of the Southern way of life."[34]

The liberal reaction to "A Letter to the North" was not quite that personal, but it was angry enough. People wanted to know about the degree of Faulkner's southern commitments; they questioned his honesty, and sometimes they were as insulting as his fellow southerners had been. "Mr. Faulkner's ancestors owned slaves," one black pointed out to *Life*, and added cuttingly, "My ancestors were slaves." This man wanted *Life* to know he was unimpressed when Faulkner advised liberals to "stop now for a moment." American blacks "have been patiently waiting a 'moment,' " he said, "which has now lasted more than ninety years." When Faulkner wrote *Life* to protest that his first statement was an "attempt . . . to save the South and the whole United States too from the blot of Miss Autherine Lucy's death," a new antagonist took up his pen. "I was in agreement with those who dismissed . . . [Faulkner's first plea] as merely an emotional, incoherent note," he said; but Faulkner's most recent letter was "a new low in groveling dishonesty."[35] Other antagonists were more famous. Aging W. E. B. Du Bois challenged Faulkner to debate the issue in the national media.[36] James Baldwin angrily denounced Faulkner's notion of moderation: "Why, if he and his enlightened confreres in the South have been boring from within to destroy segregation, do they react with such panic when the walls show any signs of falling?"[37]

Being in the middle might be important for Faulkner, but it was not pleasant. Northern and southern attitudes were polarizing, and it was fear of that polarization that had produced the most paranoid of Gavin's rantings in *Intruder in the Dust*. Now Faulkner seemed to be feeling the heat. To convince northerners like Baldwin that some southerners were indeed working for black equality, he wrote an article, "On Fear: The South in Labor," for *Harper's*, repeating his belief that being "against equality on the basis of race or color, is like living in Alaska and being against snow."[38] He overcame his hostility to the media and agreed to do a radio talk show about the crisis with Tex and Jinx McCrary; and he gave the interview to London *Times* reporter Howe, to be published in *The Reporter*. But the harder Faulkner struggled, the less he seemed to accomplish. *Harper's* did not print "On Fear" until June 1956, long after Lucy had been forced to withdraw from the university; the McCrarys, startled by Faulkner's views on integration, refused to put him on the air;[39] and Faulkner soon had reason to wish the Howe interview had suffered a similar fate.

The interview began innocuously enough.[40] "The Negroes are right," Faulkner said, "make sure you've got that—they're right"—and he wanted Howe to know that he thought blacks ought to keep pressuring southerners for equality. The Supreme Court had been right to focus on education, because both races "must be re-educated to the issue." But, as the interview continued, Faulkner seemed to change his mind about the Court's decision. Without it, he told Howe, "in fifteen years the Negroes would have had good schools" anyway, and southerners had already done a lot. "Things have been getting better slowly for a long time," he insisted, adding, "Only six Negroes were killed by whites in Mississippi last year"—he had "police figures" to prove it. That was why liberals had to be careful, he said; when things were going well, it was a mistake to speed them up. Faulkner was not quite as pessimistic about northern intervention as Gavin, who thought it would "fling . . . [the black] decades back . . . into grief and agony and violence"; he only believed the Court's "decision put the position of the Negro in the South back five years."[41]

But Faulkner's most incendiary statements in that interview had to do with his fears about polarization of northern and southern opinion. Faulkner had begun to sound polarized himself. Northern liberals did not understand what they were up against, he told Howe. "The South is armed for revolt. After the Supreme Court decision you couldn't get as much as a few rounds for a deer rifle in Mississippi." If Lucy tried to enroll again, she would be murdered. "Then the top will blow off. The Government will send in its troops and we shall be back at 1860." Faulkner could foresee a time when he might be faced with "the same choice Robert E. Lee made." His grandfather—again he confused his grandfather and his great-grandfather—had made that choice, and he was afraid he would have to make it, too: "If I have to choose between the United States government and Mississippi, then I'll choose Mississippi." He would make that choice "even if it meant going out into the street and shooting Negroes," he said. "After all, I'm not going out to shoot Mississippians." That, of course, did not mean blacks could not be Mississippians. "Ninety percent of the Negroes," he insisted, were with the white majority. "My Negro boys down on the plantation would fight against the North with me"; he would only have to say, "Go get your shotguns, boys."[42] When those words appeared a month later in *The Reporter,* an embarrassed Faulkner repudiated parts of the interview; but the pressures he had been feeling had already left their mark, and the damage to Faulkner's role as Redeemer was done.

Those pressures combined with personal problems in the cruelest fashion. Estelle's mother died. Faulkner was thrown by a horse, aggravating an old spinal injury. Miss Maud, who had suffered a stroke several months earlier, remained in a weakened condition, and Faulkner was cut to the quick when somebody told him that Miss Maud did not like the way he was using his celebrity status to air his opinions. Even chance acquaintances seemed to turn sour. Walking through the woods near Rowan Oak, he met a strange youth who wanted to know "what is it that's so different about you." Faulkner told the boy to "ask around," then "come back and tell me and I'll tell you if it's right." They met again, and the boy said, "I asked two people, and all I could find out was you're a nigger-lover." On March 18 Faulkner vomited blood and lost consciousness. If he had an ulcer, nobody could find it; before the week was out, he was telling people he was ready to leave the hospital, and he was already responding to queries about his "Go slow" statement and the Howe interview.[43] But by now it was clear that something irreversible had happened to Faulkner.

His notes to *The Reporter* and to *Time* (which had carried quotations from the Howe interview) showed what that was. Both statements were agonized and ambiguous, strained to the point of inanity to avoid outright denial of what they were written to deny. "In our troubled times over segregation," Faulkner wrote *Time,* "it is imperative that no man be saddled with opinions on the subject which he has never held and, for that reason, never expressed"; but he was careful to avoid saying what opinions he had "never held" or whether Howe had quoted statements Faulkner not only had "never held" but also had "never expressed." Faulkner protested that the "statement that I or anyone else in his right mind would choose any one state against the whole remaining Union of States, down to the ultimate price of shooting other human beings in the streets, is not only foolish but dangerous." But he did not challenge Howe's accuracy. Calling the statement "not only foolish but dangerous" might allow readers to think it was one of those opinions Faulkner had "never held" and "never expressed"; but that only denied that the statement was valid, not that Faulkner had made it. Faulkner went on to assert that Howe had attributed to him "statements . . . no sober man would make, and, it seems to me, no sane man believe"[44]—an invitation to infer that there was something about the interview that Howe had omitted and that Faulkner could not repeat publicly—that Howe was naive, and that Faulkner had been less than himself. But the reader who inferred these conclusions also had to take them for a back-handed admission that Howe had quoted Faulkner accurately. And Howe stood by the quotes: "what I set down is what he said."[45]

Unsatisfied with his notes to *Life, The Reporter,* and *Time,* Faulkner decided to go directly to a black audience, and he wrote a piece entitled "If I Were a Negro" for *Ebony.* He wanted blacks to understand what he had meant by "Go slow." He had just been trying to say, "Be flexible." What was important was getting the job done without endangering Lucy. Other blacks ought to apply to the schools now, time after time, "until at last the white man himself must recognize that there will be no peace for him until he himself has solved the dilemma." Faulkner was careful to preface these remarks with a suitable show of modesty. Like Gavin, who had told his imaginary northern antagonist, "I don't know . . . [the black] at all" and insisted "no white man does," he wanted blacks to understand he was aware of his limitations. "It is easy enough," he wrote, "to say glibly, 'If I were a Negro, I would do this or that' "; a white had to remember that he "can only imagine himself for the moment a Negro," that "he cannot be that man of another race and griefs and problems."[46]

Blacks who had heard Faulkner's Southern Historical Association address might have pondered this show of humility. There, as Faulkner praised the black American progress that had "produced Ralph Bunche and George Washington Carver and Booker T. Washington," he was already displaying an amazing ability to put his foot in his mouth. It was inspiring, he had said, that such men had sprung from a "people who only three hundred years ago were eating rotten elephant and hippo meat in African rain-forests, who lived beside one of the biggest bodies of inland water on earth and never thought of a sail, who yearly had to move by whole villages and tribes from famine and pestilence and human enemies without once thinking of a wheel."[47] Blacks who heard that might have been excused if they were unenthusiastic to hear from Faulkner that they had barely escaped savagery by being brought to America, that their ancestors had subsisted on a diet of carrion, that they had been too dull to invent wheels and sails. And if Faulkner did not know much about Africa, he did not seem to know much about black Americans, either. Carver and Washington, after all, were nobody's Black Panthers, and if Faulkner thought Du Bois was a distinguished American, he was not talking about it.

Now, in "If I Were a Negro," Faulkner displayed similar problems. If his notion of "flexibility" was all he had been trying to express in that essay, few blacks might have objected; many advocated the same approach. But Faulkner had other views that he wanted to express to *Ebony*'s readers. The "founding fathers," he said, had intended people to have "the right to *opportunity* to be free and equal," but they had only intended that right "provided one is worthy of it, will work to gain it and then work to keep it." There were "the responsibilities of that opportunity," Faulkner said, and he had a formidable list of those. They included "the responsibilities of physical cleanliness and of moral rectitude, of a conscience capable of choosing between right and wrong and a will capable of obeying it, of reliability toward other men, the pride of independence of charity or relief." It was the black's "tragedy," Faulkner said, "that these virtues of responsibility are the white man's virtues of which he boasts, yet . . . the Negro, must be his superior in them." If he were black, he would tell fellow blacks, "We as a race must lift ourselves by our own bootstraps to where we are competent for the responsibilities of equality, so that we can hold on to it when we get it."[48]

Faulkner had come full cycle. Flying around the world for the State Department and urging Americans and southerners to clean their own house so they could face the Cold War, he had hoped to help resolve his nation's divided loyalties; now, under the pressure of "being in the middle," he was finding he could not cope with the split in his own, and one by one his deeper commitments were surfacing. Since the Howe interview, no one doubted what his feelings on northern interference were; now he had let slip his notions of black character. Faulkner had often hinted that his beliefs on that subject were like Isaac McCaslin's; now he sounded closer to Cass Edmonds, who thought black vices were inherent, incurable ones: "Promiscuity. Violence. Instability and lack of control. Inability to distinguish between mine and thine."

IV

Miss Maud had been right about the power of celebrities to focus public attention. Their success in one field argued the authority of their efforts in others, so that any commitment by a celebrity—even a confused, ambiguous commitment—appeared important. For Faulkner, that was a considerable responsibility, and, whatever his failures, he had struggled hard to live up to it. Still, what was striking about Faulkner's efforts to be "articulate in the national voice" was his ineptness. His inconsistencies confused many admirers, and, if the anger those efforts inspired was any gauge, they also hardened people on both sides in their extremism. But if Faulkner's stand seemed to have had little effect on those for whom it was intended, its effect on him was clear.

Faulkner's campaign for moderation was never the same after he was hospitalized. Faulkner had called on blacks to be "flexible," but he was finding it more difficult to be flexible himself. It was not so much that his views were changing; they were just subtly hardening. He had few kind words for the die-hards in Mississippi. "The liberal in Mississippi" was "hemmed in . . . by the sort of people that burn crosses and whip Negroes and compose lynching parties," he said in 1958; there was a "noisy majority"—then he changed that to "minority"—of Mississippians who "if you say something or do something they don't like they may set your house on fire." Sometimes he sounded just as hostile to the northern liberals who were standing up to such racists. Like Gavin, who wanted northerners to "come down here and look at us," Faulkner insisted that "if they would come down here and look at . . . [the race problem], they would know a good deal more than they do now." Northern liberals were "going to learn" anyway, "because there are more and more Negroes moving to the North."[49]

But the barometer of the hardening of Faulkner's attitudes was a new belligerence about black character. In a question-and-answer session after his 1958 speech, "A Word to Virginians," he was talking more openly than in "If I Were a Negro." Although admitting that "the white man is responsible for the Negro's condition," he seemed to be blaming blacks for that condition anyway. It was a "fact," he said, "that the Negro does act like a Negro and can live among us and be irresponsible."[50] In a speech that preceded that statement he had been almost as direct. "For the sake of the argument," he asked his Virginia audience, "let us agree that as yet the Negro is incapable of equality"; he wanted them to agree that this was true "for the reason that . . . [the black] could not hold . . . [freedom] and keep it even if it were forced on him with bayonets" because, "once the bayonets were removed, the first smart and ruthless man, black or white, who came along would take it away from him." Granted such agreement, Faulkner thought, it followed that white men had to "take . . . [the Negro] in hand and teach him" what Faulkner called "the responsibilities of equality," and he was ready to spell out what those were. "What he must learn are the hard things—self-restraint, honesty, dependability, purity." What all that meant to Faulkner was that the black "must learn to cease forever more thinking like a Negro and acting like a Negro" —and Faulkner added somewhat superfluously, "This will not be easy for him."[51] Even that was not the strangest part of this performance. "Perhaps," Faulkner speculated, "the Negro is not yet capable of more than second-class citizenship"—and suddenly he seemed to be giving something away. Through a long career Faulkner had had every bright, successful black in his novels except Dilsey and Ringo attributing his success to a non-African heritage. Now he said that the black's "tragedy may be that so far he is competent for equality only in the ratio of his white blood."[52] The mask was off. Faulkner's opinion of the majority of blacks had never been much different from that of Cass Edmonds.

Isaac and Chick had been prepared for social commitment by the inspiration they drew from "primitives." Isaac's conflicts about his repudiation of the plantation had been resolved by the wilderness mysticism he had learned from Sam Fathers; Chick's heroism had been motivated by the strength he had learned from Lucas Beauchamp. But Sam and Lucas were the very creations that had allowed Faulkner to avoid the full impact of the South's guilty past; they had allowed him, under the guise of exploring the ways their "primitive" virtues could inspire his white heroes, to shield those heroes from the realities of black experience. Now Faulkner had thrust himself into a position not unlike that of his heroes; but, when he looked back into his own psyche for the inspiration that had guided Isaac and Chick, he found only the divided allegiances of the world of his youth, preserved through the years under the "primitive" masks he had projected onto his black heroes and heroines—and those allegiances produced no unified commitment to the moderate position, but instead a kind of social schizophrenia. Faulkner's efforts showed that he was not the man who can say "This stinks and I'm going to do something about it." He was no Chick Mallison, daring existential action in the face of blind unreason and failed communication. He was not even an Isaac McCaslin, swearing "at least I will not participate in it myself." He was still very much the author of *Absalom, Absalom!* and *Go Down, Moses,* who did not know that he had failed to face the full truth about the South's guilty heritage, and the author of *Intruder in the Dust,* who, because of that failure, could portray heroism only in terms of fantasy and sensationalism. The divided allegiances that had insured the failure of his novels had also insured his failure as a political voice.

But, if Faulkner's vision of a redeemed South had, finally, gone the way of all of his other visions of the South, there were compensations. There was another South beckoning—and it did not seem so very far away. Family tradition had the first Falkners in America landing at Charleston; the Sartorises, Compsons, and McCaslins all looked back to Carolina ancestors; no aristocratic family that came from Virginia played a significant role in his works.[53] But by 1958 Faulkner was discovering Virginia. Jill Faulkner and her husband, Paul D. Summers III, had been living near Charlottesville since 1956, and Jill had given birth to a son, Paul IV. Faulkner, partly to be near them, had accepted a term as writer in residence at the University of Virginia. What followed was remarkable. Faulkner seemed at home in Thomas Jefferson's university, and in "A Word to Virginians" he seemed to be trying to articulate what Virginia had come to mean to him. Virginia was "the mother of all the rest of us of the South," he said. Since "the Virginia younger sons" had "moved . . . with the surplus slaves two hundred years ago" into "the other states of the South," there was "a definite tie." This was a tie "of blood, of culture, of thought, of having shared in the same disasters." For those reasons, he thought, "the soul of Virginia feels the same tie toward Mississippi and Alabama and Georgia as the soul of Mississippi and Alabama feel of respect toward the mother stock in Virginia." It was because these ties were so strong that he was making his appeal to Virginians for leadership in the present crisis. Virginians ought to feel a paternal relationship toward other southern states, "something of the same responsibility . . . that the father might feel toward his sons." If Virginians faced up to integration, other states might, too. "Show us the way," he pleaded. "I believe we will follow you."[54]

That appeal showed where Faulkner's search for a South was taking him as his career wound down. It was to Virginians, not Mississippians, that he was making his plea, and that suggested that it was Virginia that had come to stand for those older values that he had associated with the world of his youth. Four months later his answer to a question about how he liked Virginians was suggestive. "The longer I stay in Albemarle County the more like Mississippians they behave," he said, "but they don't have some of the Mississippi vices." That sounded like a breakthrough for Faulkner: he had learned to project his notion of aristocrats onto Virginians and his notion of rednecks onto Mississippians. Faulkner had spotted "a few Snopeses" in Albemarle County, he said, but that did not bother him. "I said last year that I liked Virginians because they were snobs, and since they seem to think I'm all right then maybe I am all right."[55]

Virginia had taught Faulkner a significant lesson: it had shown him, at last, how to separate his memories of the world of his youth from their physical location; and having learned that, he had discovered that he could love that world without feeling responsible for its faults. After December 1958 Virginia had still another appeal: Jill had given birth to William Cuthbert Faulkner Summers. By September 1959 Faulkner was still dividing his time between Virginia and Mississippi, but he had bought a fine Georgian home on Rugby Road in Charlottesville. To this point Faulkner's ventures as the South's new Redeemer seemed to have accomplished little. The regional and national housecleaning that would allow Americans to present a united front seemed a long way off; both sides seemed to be marshalling for what promised, at the least, to be an era of bloody riots in his homeland, and Faulkner had reason to believe, as he had told Howe, that the country might soon "be back at 1860." Faulkner had struggled hard in the years since he had told Jill's class never to be "afraid to raise your voice . . . against injustice and lying and greed"; he had pursued that crusade even to the point of risking his health. But by 1959 he had every reason to hope that people would judge his political efforts the way he wanted them to judge his novels: on the "courage" he had shown taking chances rather than on whether he had succeeded in reaching his goals.

By the mid-1950s Faulkner had developed a standard answer to questions about what he had achieved. He "rated" his contemporaries on "the magnificence, the splendor, of the failure," he said. He thought "the writer" had to struggle for "perfection," and such a struggle predestined him to failure: "He can't do it in his life because he can't be as brave as he wishes he might, he can't always be as honest as he wishes he might." But a good writer never gave up on his dream; he could always "hope that . . . he can make something as perfect as he dreamed it to be."[56] For Faulkner's efforts as Redeemer, at least, that was the best that could be said. But by 1959, playing with his grandsons on Rugby Road, he seemed to have his mind on a different kind of perfection.

Two years earlier Faulkner had still been insisting that "it's foolish to be against progress"; but, as he reminisced about "the sort of background which a country boy like me had," the South he was about to find in Charlottesville began to surface. Faulkner thought it was "a sad and tragic thing for the old days, the old times, to go." He was sure everyone felt that way. "I don't want it to change," he said. When a man got old, he saw life differently. "Probably anyone remembers with something of nostalgia . . . his younger years. He forgets the unpleasant, the unhappy things that happened, he remembers only the nice things."[57]

Faulkner's new mood did not mean that he had quit worrying about what he could do to protect his homeland; it was just that his part of the fight was over. Faulkner, like most men, might not always have been "as brave" or "as honest" as he had set out to be, but he had done his best, by the lights he had inherited from the world of his youth, to be responsible and protective. He had brought the official standards of the Old Order to this new Reconstruction; he had set his course directly toward the danger his tradition told him threatened his homeland, and, now that he had failed to forestall the inevitable, he was free. At last he could lash out at what frightened him in his homeland without feeling that he was failing it; and he could love it freely, without feeling that by loving it too well he was failing his conscience. It was not so much that Faulkner was resisting change; he had simply passed beyond it. He had made his peace with change, and now he was impervious to it.

12

Home

When fears of aging began to trouble William Faulkner in his mid-forties, he spoke of a "period" artists reached in their "early fifties," a time when "they admit that at last there is no solution to life and that it is not, and perhaps never was, worth the living." Faulkner had consoled himself by speculating that after the war "perhaps the time of the older men will come." These men would not yet be "so old that we too have become another batch of decrepit old men looking stubbornly backward at a point 25 or 50 years in the past"; their "time" would have come because they "have been vocal long enough to be listened to." By 1961 few Americans would have ventured that Faulkner's "time" had already passed; but Faulkner found that old feeling that there might be "no solution to life" reasserting itself.

Shakespeare was on his mind. With *The Tempest,* he thought, Shakespeare had "said at last, 'I don't know the answer, so I will break my pencil and stop.' " That seemed to move Faulkner. "At my age I know how he felt," he said, and added with teasing ambiguity, "You never will have the answer to the human condition so you might as well give up."[1] Faulkner sounded almost relieved, as though that was what he wanted to do; but whatever "answer" he was talking about, he thought he had learned others. "I like to think anyone grows as he gets older," he said. "He may not have the power and drive he had at 20, but he prefers to believe he understands more." Age made a man more tolerant. "He's not always able to forgive human folly, but he is able to understand it."[2]

That in itself was a kind of "answer to the human condition"; if Faulkner could "understand" the "human folly" he saw around him, it might not be quite so important to identify what "stinks" and to "do something about it." Finishing off the *Trilogy of Snopes* he had begun in 1926 as *Father Abraham,* he had still been firing Parthian shots in his battle to preserve the moderate position. In *The Mansion* (1959) he seemed to be going out of his way to show that southerners were capable of cleaning their own house. Snopes-antagonist V. K. Ratliffe set the example by persuading two boys at a rally to smear demagogue Clarence Snopes's trousers with brush from a "dog post office"—with the result that Snopes was chased from the rostrum and out of politics by a gathering of urinating dogs.[3] In a more practical vein, Faulkner used Flem Snopes's stepdaughter Linda, just back from Greenwich Village, to show why liberals ought to "Go slow." Impatient, Linda wanted a program to allow Jefferson's black students, held back by old-style black teachers, to work with more demanding white ones; soon Faulkner's black principal was asking Gavin Stevens to keep Linda from interfering. The best thing for blacks, this principal said, was to make themselves so important to the economy that whites could not do without them; but until then they did not want help from patronizing liberals. "Just say we thank her and we wont forget this. But to leave us alone" (*TM*, 225).

Faulkner was finding it hard to quit charging the barricades. But for all his militance, a feeling that his "time" might have already passed seemed to be creeping into his trilogy. Faulkner seemed obsessed with visiting old friends among his characters, embellishing old tales. Even Flem Snopes now seemed less threatful than disgusting. Mink Snopes, Flem's aging nemesis, emerged from a thirty-eight-year incarceration in Parchman prison to avenge himself on the cousin who had put him there; and, as he walked out under the stars after killing his old enemy, it was impossible not to connect sixty-three-year-old Mink with sixty-two-year-old Faulkner. Exhausted, Mink felt an overpowering desire to lie down and sleep. He reflected that "the ground . . . never let a man forget it was there waiting"; he could feel it "pulling gently and without no hurry at him between every step." It said, "Come on, lay down; I aint going to hurt you. Jest lay down" (*TM*, 434).

As Mink slept under the stars, Faulkner found that his search for a South had taken him to a familiar place. For all his reluctance to give up his struggle to forestall the new Reconstruction he feared was imminent, that struggle was over for Faulkner; what remained for him was his mood of mellow nostalgia. It was this new mood that showed Faulkner where he was. It, too, was a kind of "answer to the human condition," and the novel he now wrote showed what this mood had meant to him. *The Reivers* (1962) was set in 1905, exactly seven years further in the past than he had once allotted those "decrepit old men looking stubbornly backward," but Faulkner seemed to have a marvelous time writing the novel. "This book gets funnier and funnier all the time," he told a friend,[4] and when he finished he subtitled it, *A Reminiscence.* Faulkner was ready for one last mellow romp through the world of his youth.

As aging Lucius Priest told his grandson about that world, it had a familiar ring. Lucius had been eleven in 1905, three years older than Faulkner had been that year. Like Faulkner, Lucius had been named for a distinguished great-grandfather; and his grandfather "Boss" Priest, like J. W. T. Falkner, was an old-style aristocrat who ran the county's politics. The Winton Flyer Boss owned sounded like the Young Colonel's Buick, and even the mud hole it got stuck in at Hell Creek bottom resembled one that had given the Falkners trouble.[5] Like J. W. T. Falkner, "Boss" seemed to dominate everything in his family, including Lucius's less charismatic father, Mr. Maury Priest who, like Mr. Murry Falkner, ran a livery stable. Lucius had been raised by a bustling, razor-tongued black woman named Mammy Callie, and Ned McCaslin, the black companion on Lucius's adventures in the novel, behaved even more like Ned Barnett than Lucas Beauchamp had. Faulkner was revisiting not only his family but old friends among his characters, too. From *Go Down, Moses* he brought back the bumbling, part-Indian Boon Hogganbeck; and Miss Reba, the beer-swilling madame from *Sanctuary*, was back with all her girls. But if the passenger list for this new tour through the world of Faulkner's youth was familiar, the relationships among the passengers had been subtly altered.

Once again, Faulkner's hero was a young aristocrat on the threshold of the adult world, and again his guides through the rites of passage were an older white from his own family and a black with patrician ancestry. But Lucius, a cousin of Isaac McCaslin, was a year younger than Isaac had been when Sam Fathers marked him with the blood of his first buck, a year younger than Chick Mallison when Lucas Beauchamp helped him from the creek; and at eleven, Lucius was more knowledgeable in the traditions of his class than either Isaac or Chick had been. Ned McCaslin, moreover, another black grandson of old Carothers, was more aggressive than either Lucas or Sam and more at home with other blacks; and Boss Priest was older and more experienced than either Cass Edmonds or Gavin Stevens had been. To this familiar trio Faulkner added a new figure. His wilderness days behind him, Boon identified himself as a poor white and was working at the Priest livery stable; Faulkner could thus have all three classes of Yoknapatawpha society participating in Lucius's initiation.[6]

The tale began with an aging Lucius trying to tell his grandson what it meant to be an aristocrat. While growing up, Lucius himself had learned the meaning of the word the hard way when his father and mother had gone with Boss to Bay St. Louis. Without those adults around, he had fallen easy prey to what he called "non-virtue" (*TR*, 54), which had confronted him in the forms of Boon and Ned; Boon persuaded Lucius to steal Boss's Winton Flyer for a trip to Memphis, and Ned stowed away under a tarpaulin in the back seat. In Memphis these two adult ne'er-do-wells quickly made Lucius conscious of what the stakes in this escapade were; Boon headed straight for Miss Reba's, and Ned seemed to be going on a rampage. Hoping to buy his cousin Bobo Beauchamp out of debt, Ned arrogantly swapped the Winton for Lightning, a race horse Bobo had stolen from his employer, Mr. Van Tosch; convinced he could make Lightning run faster than anyone so far had been able to, Ned meant to race him in nearby Parsham, pay Bobo's debts with the winnings, then swap Lightning back for the car. Appalled, Lucius saw that his acquiescence in "non-virtue" was getting him into trouble beyond his wildest dreams. But that trouble was leading Lucius toward something else as well; as the three conspirators rambled around north Mississippi and west Tennessee, he was also getting a picaresque education in the class structure of the South of 1905. For beneath the mellow nostalgia of Faulkner's new novel was a not-so-mellow polemic: a final rationale of the political and social attitudes of the Falkner family just after the turn of the century.

At eleven, Lucius had already been indoctrinated in the ethics of what he called "our Jefferson leisure class" and in particular into its neo-feudal structures of *entendre de noblesse* and *noblesse oblige.* After his initial error in helping steal the car, Lucius became something of an *enfant terrible* of *noblesse.* Patrician manners were better than a credit card in Faulkner's 1905 South. Everywhere, people recognized Lucius's breeding and accepted his status; Ned and Boon were protective conspirators; Miss Reba and her girls assumed an almost parental attitude, struggling to keep profanity from their lips. When eleven-year-old Lucius had no past experience to draw on, he found his awareness of *noblesse oblige* stood him in good stead. He thrashed fifteen-year-old Otis, nephew of Boon's prostitute girl friend Corrie, for running a peep show with Corrie as the unwitting star; and to reward her young knight, this grateful whore, much to Boon's discomfort, promised to reform. Like the responsible aristocrat he was, the boy loyally stuck with Boon and Ned, both of whom were his family's dependents, shouldering the responsibility of riding Lightning in the race himself; and, when he had finished, Lucius refused the money from the bet Miss Reba had made for him. "I wasn't doing it for money," he insisted; it was just "that once we were in it, I had to go on, finish it." Lucius's virtuousness did not end with horses, either; he was a paragon of patrician sensitivity to the race problem. Calling people "nigger," Lucius knew, was gross. "Father told me," he insisted, "that a gentleman never refers to anyone by his race or religion." In Parsham, as guest of black Uncle Parsham Hood, Lucius was offered a bed belonging to the old man's grandson; but Lucius cut quickly through the problem that had brought a curse on Roth Edmonds in *Go Down, Moses.* Lucius wanted to sleep with Uncle Parsham, he insisted, and he even came up with a joke to get his black host through the awkward moment. "I sleep with Boss a lot of times," he said. "He snores too. I dont mind" (*TR,* 37, 279, 143, 250).

In one sense Lucius was negating his patrician breeding by making this wild trip; in a deeper sense he was giving evidence of that breeding. Lucius was not being initiated into "non-virtue"; he was learning how his inherited class virtues allowed a man to deal with "non-virtue," and in *The Reivers,* at least, that took care of everything. Faulkner seemed to have forgotten his fears about the South's guilty past. Lucius never had to face the lonely confrontation with inherited guilt that Isaac McCaslin faced in *Go Down, Moses.* His problem was the opposite of Isaac's; when Boss Priest unexpectedly showed up in Parsham in the middle of the race, Lucius had to face the consequences of his own, not of his ancestors', misdeeds. Lucius never feared, as Isaac did, that his class might be corrupt; he was afraid that he himself was and that his class would never accept him again. "I lied," he told Boss later, and insisted that his grandfather "do something about it. Do anything, just so it's something." When Mr. Maury, his father, wanted to settle things by whipping him, Lucius knew immediately that that was not going to work, that "if after all the lying and deceiving and disobeying and conniving I had done, all he could do about it was to whip me, then Father was not good enough for me" (*TR,* 299, 301). Fortunately, wizened Boss intervened, taking the boy aside for a solitary talk that succeeded where a whipping never could have.

The problem, Boss told Lucius, was not that Lucius himself had instigated the actions he regretted; Boon and Ned had done that. The problem was that Lucius had "acquiesced to them, didn't say No though he knew he should." But the patrician ethic, Boss showed the boy, was viable precisely because it showed aristocrats how to deal with their own aberrations. Innocent Isaac McCaslin, repelled by the guilt of his class, had been driven into tragic isolation; Lucius learned how to make a virtue of his own guilt and that reconciled him to his class. "Live with it," Boss told him. "Nothing is ever lost. It's too valuable." *Noblesse oblige* extended to one's self as well as to others. "A gentleman accepts the responsibility of his actions and bears the burden of their consequences," Boss said. "A gentleman can live through anything. He faces anything" (*TR,* 302).

Such was the real meaning of Lucius's round with "non-virtue": he could rejoin his family stronger now because he knew how to accept responsibility for himself. But the people responsible for "non-virtue" in Lucius's world were usually plebeians, and the aristocrat, knowledgeable in matters where they were not, had to minister to their shortcomings as well as his own. At that point *The Reivers* became an allegory of the social virtues of 1905 Mississippi, surveying all the classes from the standpoint of *entendre de noblesse* and *noblesse oblige.* Faulkner's mellow nostalgia allowed him a tolerant view of both black and white plebeian classes in that world.[7] No Snopes appeared in *The Reivers*; the Puritan fanaticism of *Light in August,* even the gruff vengefulness of the Beat Four hill clans of *Intruder in the Dust,* seemed remote. Miss Reba and her girls were kindly whores; railroad man Sam Caldwell was dependable and unprejudiced; Mr. Poleymus, the Parsham constable, exuded integrity.

But for all that, *The Reivers* showed, it was still the poor white class that, in Faulkner's world, provided the troublemakers. It was the poor-white boy Otis who prompted Lucius's observation that "a gentleman never refers to anyone by his race or religion" and who flew into a rage at Ned, "cursing Ned, calling him nigger"; and it was Parsham's poor-white sheriff Butch Lovemaiden who emerged as the novel's central image of what "non-virtue" meant. Butch bullied Uncle Parsham Hood, calling the old man "boy"; he lorded it over Ned and white Boon, too, secure in the knowledge that they had to put up with him for the sake of "Uncle Parsham's home and family." True to his class, Lucius felt somehow responsible for all that. "I was ashamed," he recalled, "that such a reason for fearing for Uncle Parsham, who had to live here, existed" (*TR,* 143, 172, 174).

Aristocrats had to control monsters like Butch for the sake of other plebeians; but other whites had to be looked after for their own sakes. Part-Indian Boon, who identified himself in this novel as a poor white, was a case in point. "Boon didn't actually belong to us," Lucius explained; he also belonged to other patricians, "to Major de Spain and General Compson too"—they all remembered Boon's role in the old days in the wilderness and felt a family obligation to him. But Boon, a pathetically ineffective hunter, was little better at anything else, and the real stable work was done by blacks; Boon "was a mutual benevolent protective benefit association," Lucius recalled, "of which the benefits were all Boon's and . . . the benevolence and the protecting all ours" (*TR,* 18, 19). Since irresponsibility was not reserved for blacks, the aristocrat's duties were extended laterally to Jefferson's other plebeian class.

What went on among Mr. Maury, Boon, and the black stable hands showed how *entendre de noblesse* and *noblesse oblige* worked. A black driver, Ludus, knew how to take advantage of *noblesse oblige*: he kept a team out all night while visiting his girl, secure in the knowledge that this offense, though serious, was not going to cost him his job. As Ludus's white man, Mr. Maury was responsible for his subsistence; Mr. Maury's problem was that, without severing the relationship, he had to find a punishment strong enough for the offense. The expedient he chose was possible only under the carefully structured traditions of segregated Mississippi: he fired Ludus without paying him his weekly wage. Ludus understood what Mr. Maury had done. Firing him was Mr. Maury's way of dramatizing how serious his offense was; but Ludus also knew that by refusing his pay, Mr. Maury had broken a patrician obligation and was obliged to compensate him. That, Lucius explained, was Mr. Maury's signal to Ludus that he was not fired after all. Rather, it "indicated that he was merely being docked a week's pay (with vacation) . . . ; next Monday morning Ludus would appear with the other drivers at the regular time and [black foreman] John Powell would have his team ready for him as if nothing had happened" (*TR*, 14).

John Powell was more responsible; he understood *entendre de noblesse* as well as *noblesse oblige.* John was a dependable worker, but he had an outrageous eccentricity; there was a "decree," Lucius related, "as old as the stable itself, . . . that the only pistol connected with it would be the one which stayed in . . . the desk in the office." As a farm youth, John had worked overtime and saved his money until, at twenty-one, he could buy a pistol from his father; he then told the other stable hands that pistol was "the ineffaceable proof that he was . . . twenty-one and a man." John "declined even to imagine the circumstance in which he would ever, pull its trigger against a human being." Still, Lucius insisted, John "would have no more left the pistol at home . . . than he would have left his manhood in a distant closet" (*TR*, 6, 7). John's pistol was his eccentricity, and whites put up with it because they liked him.

To nobody's surprise, it was poor-white Boon who breeched that decorum. Ludus, he suddenly proclaimed, had wronged him; he had not only sold Boon bad whiskey, he had told another black that Boon "was a narrow-asted son of a bitch." Boon's status as a white was at stake, and in his anger he stole John's pistol. "I'm going to shoot Ludus," he told the appalled Mr. Maury. Pathetically unable to hit anything with a gun, Boon fired at Ludus and instead grazed the hip of a nearby black girl—and "the whole edifice of *entendre de noblesse* collapsed into dust." Everybody in town was going to know about John's pistol, and Mr. Maury would have to get rid of John or persuade John to get rid of the pistol. But "the *noblesse*, the *oblige*," Lucius pointed out, "still remained" for Mr. Maury, who had to look out after Boon and Ludus. Those two were little help, and their argument continued in patrician Judge

Stevens's court. When Boon insisted Ludus had "insulted" him, Ludus responded outrageously, "I never said he was norrer-asted," he replied. "I said he was norrer-headed." Boon could hardly bear that. "Me, a white man," he lamented to the judge, "have got to stand here and let a damn mule-wrestling nigger either criticize my private tail, or state before five public witnesses that I aint got any sense." That remark showed Boon's plight as a poor white; he received no respect from blacks that he could not earn. "He was almost crying now," Lucius recalled (*TR*, 15, 5, 9, 16).

But with the law's help Mr. Maury managed to solve the problem. Sheriff Hampton's first thought was the injured black girl. Since the shot "barely creased her," he suggested that Mr. Maury ought to "buy her a new dress . . . and a bag of candy and give her father ten dollars"—that, he was sure, would "settle Boon with her." Mr. Maury improvised a solution for Boon's and Ludus's quarrel. "I will make the bond," he told Judge Stevens. "Only, I want two mutual double-action bonds," each of them to "be abrogated, . . . at the same moment that either one of . . . [the two offenders] does anything." That way Ludus and Boon had to get along with each other, because if either took up the feud again, both had to pay the price of the bond to Mr. Maury in work. That strange artifice was a paradigm of the aristocrat's solution for the race problem: it was the South's two plebeian classes who created the friction, and, if they were going to keep doing that, they ought to pay the price in bondage. "I dont know if that is legal or not," Mr. Maury told Judge Stevens, who answered, "We can try. If such a bond is not legal, it ought to be" (*TR*, 15, 16). If it was outrageously *il*legal, for the moment that seemed beside the point; what was important to Faulkner's "Jefferson leisure class" was keeping the power in the hands of people who had to take the responsibility. Such was the reality of *noblesse oblige.*

But for aging Lucius Priest, the ethic of his youth was more than *noblesse oblige* and *entendre de noblesse*; he wanted his grandson to know patricians held a genuine affection for their black charges. Faulkner's attitude toward blacks in *The Reivers* had subtly changed. If an aging writer "understands more," as he claimed, that seemed to have a lot to do with all that had happened in the thirty-six years since *Soldiers' Pay.* That "Homer of the cotton fields" who in *Sartoris* had found similarities between blacks and mules might still be hovering in the wings, but in the mellow nostalgia of *The Reivers* he had undergone a transformation. The novel's two central black characters, Uncle Parsham Hood and Ned McCaslin, were designed as sensitive tributes to their race; and, although the subtlety of Faulkner's more relaxed approach did not cover his blind spots, it hid them better.

If these strong, talented black men seemed familiar, it was no accident;

like Ringo Strother and Lucas Beauchamp, they had Ned Barnett as a common ancestor. This time Faulkner split Uncle Ned's personality two ways: Ned McCaslin had inherited Uncle Ned's wit, drive, and scheming intelligence, and Uncle Parsham, his dignity and strength. Lucius was impressed with Uncle Parsham, who lived in fruitful retirement amid the respect of family and peers. He was "very dark" complected and seemed "even regal," Lucius recalled, "prince and martinet in the dignity of solvent and workless age." Uncle Parsham's evening meal set the tone in his household. He spoke to his maker "without abasement or cringing," as "one man of decency and intelligence to another." Uncle Parsham was not going to let God take him for granted. He was "notifying Heaven that we were about to eat and thanking It for the privilege," but he was sending out another message as well: he was "at the same time reminding It that It had had some help too" (*TR*, 167, 224, 247).

If Uncle Parsham Hood looked and acted black, Lucius recalled, it was worth remembering that the town he lived in was named Parsham, too; he "bore in his Christian name the patronymic of the very land we stood on." Whether or not he was kin to any white Parshams, the old man's style defined his roots, recalling Ned Barnett's affinity for the Falkners' clothing and mannerisms. Lucius saw Uncle Parsham first in "a white shirt and galluses and a planter's hat." At the table he "unfolded his napkin and stuck the corner in his collar exactly as grandfather did," and, afterward, he sat "picking his teeth with a gold toothpick just like grandfather's." Wherever he had inherited those mannerisms, to Lucius, at least, Uncle Parsham was "the patrician" of the strange group he had joined, "the aristocrat of us all and the judge of us all" (*TR*, 176, 167, 246, 247). This mature, dignified man showed Lucius that the patrician code was viable for blacks, too, and that the black who observed its nuances could live a rewarding life in a partitioned society.

It was those patrician manners of Uncle Parsham, finally, that defined the problem Faulkner faced in his reminiscence. Faulkner still did not seem to see how such a characterization cut two ways; Uncle Parsham's skin might be black, but whether or not he was kin to any white Parshams, the heritage he lived by was white. That "understanding" Faulkner thought older writers gained might be new to him in *The Reivers*, but what he was doing with Uncle Parsham was familiar. Every defense of the South's aristocracy since John Pendleton Kennedy's *Swallow Barn* had featured suitably grateful blacks as witnesses to the plantation's beneficence; Faulkner had been producing such blacks since *Soldiers' Pay*, and the benign picture of 1905 Mississippi patricians he was painting meant he had to do it again. And what was true of Uncle Parsham was true of the novel's major black figure, Ned McCaslin.

Even more than Uncle Parsham, Ned McCaslin recalled Ned Barnett. Like Ned Barnett he had his comic side, spouting malapropisms in dialect and punctuating them with a raucous "hee hee hee." He was shiftless enough to satisfy the most die-hard segregationist. He was a tomcat who had had four wives and made a pass at every black woman he could reach. Ned could be gross, too, as Lucius and Boon discovered when he betrayed his hiding place under the tarpaulin by passing gas. His gaucheries made him the perfect foil for civilized, protective Lucius; so did his old-fashioned attitude toward his white family, and here Ned resembled part-white Lucas Beauchamp more than Ned Barnett. Ned McCaslin "was our family skeleton," Lucius reported, "a McCaslin, born in the McCaslin back yard"; the Priests had "inherited him" along "with his legend," which was "that his mother had been the natural daughter of old Lucius Quintus Carothers himself and a Negro slave." Having a white ancestor seemed to mean as much to Ned as it had meant to Lucas; and, as with Lucas, his identification with this ancestor was the signal of a regressive quality in his self-image. His sartorial habits showed it; these were as bizarre, and as old-fashioned, as Ned Barnett's. For the Memphis trip he wore his Sunday outfit, which included "a black suit and hat and . . . white shirt with . . . [a] gold collar stud" but "without either collar or tie." He carried a "small battered hand grip" that old Carothers had owned; it contained, along with an almost-empty whiskey flask, a Bible that had belonged to Carothers's wife (*TR*, 30–31, 69). Ned was as regressive in his white identification as he was in his black; like Lucas Beauchamp, he identified himself not merely with old Carothers but with his ancestor's generation.

The white aspect of Ned's self-image pervaded his social attitudes. As the leader of the expedition to Parsham, he practiced his own notions of *noblesse oblige.* He was always on hand to look after Lucius; he made a risky but unsuccessful attempt to separate Corrie from the lusting Butch; even the horse-racing scheme was an effort to protect his cousin Bobo Beauchamp. What was new about Ned was his identification with blacks, and that made the difference. Dilsey Gibson, Ringo Strother, and Charles Bon had rejected their black peers to identify with white families, while Sam Fathers and Lucas Beauchamp had retired into proud isolation, their tragic lives suggesting the curse that white greed had brought on the land. But Ned was the first of Faulkner's blacks to participate fully in the black life around him. If there was any curse on his patrimony, Ned seemed blissfully unaware of it, and his freedom identified the new mood Faulkner had brought to *The Reivers.*

Electrically aggressive, Ned overcame every handicap blacks suffered in a partitioned society. Joe Christmas had been destroyed by color barriers; Ned made them his tools. Roth Edmonds had observed Lucas Beauchamp becoming "not Negro but nigger," and Ned, Lucius observed, could play "Uncle Remus" when it suited him; but that, Lucius noticed, was never "when it was just me and members of his own race around." Around Lucius Ned's demeanor was an "avuncular bossiness"; around Boon it was a "spoiled immune privileged-retainer impudence"; around Miss Reba on a matter of business, those attitudes were "completely gone." Whatever the role, he manipulated everybody. Acutely aware of his dependence on Boss Priest, Ned showed a crack in his armor sometimes; when Boss surprised him at Parsham, Lucius reported, Ned "looked like he had without warning confronted Doom itself." More often, Ned was quietly expert in his relationship with his patron. When Lucius asked Ned why he had not helped Boss recoup his losses in the last race, Ned was perceptive. "When I offers to pay his gambling debt, aint I telling him to his face he aint got enough sense to bet on horses? And when I tells him where the money came from I'm gonter pay it with, aint I proving it?" (*TR*, 182, 128, 270, 304). Ned never had to transcend the color line; he beat it to death with its own ambiguities. The new freedom Faulkner had been feeling permeated Ned's portrait, with the result that his final black character came through as the most dominating, liberated black he had created. Everything Faulkner had learned and written about blacks came together to produce Ned. There was something of Dilsey's grumpy loyalty, of Ringo Strother's free-wheeling assertiveness, and of Lucas Beauchamp's manipulatory deviousness; and it was all fused to produce a figure that was finer than the sum of its parts. The irony was that, fine as his last black creation was, Faulkner had enlisted him in the same cause in which he had enlisted his others.

In a final confrontation with Boss Priest, Faulkner brought Ned forth as an articulate defender of the segregated paternalism of 1905 Mississippi. The atmosphere for this tongue-in-cheek scene was heavy with moonlight and magnolias. The setting was old Parsham Place, home of Colonel Linscomb, who owned the horse Lightning had run against. Lucius recalled that it was "big, with columns and porticoes and formal gardens," and he remembered seeing "what used to be slave quarters." They sat by a window; there was the odor of "roses . . . , and honeysuckle too"; there was "a mockingbird somewhere outside." Colonel Linscomb was in white linen, and Boss in his Confederate gray. Toddies were served to all the adults, including Ned, who was priming himself to cover for his misadventures with the performance of his life. Politic to a fault, Ned declined to drink until told to, ignored a second when it was poured. Refusing to look at Boss, he began to unravel the tale of Bobo's difficulties; he guided the conversation toward Colonel Linscomb's other visitor,

Bobo's former employer Mr. van Tosch, a Chicagoan who had settled in Memphis to breed horses. Soon Ned's plea for Bobo became an initiation of this outlander—"an alien, a foreigner," Lucius recalled—into the ways of *entendre de noblesse* and *noblesse oblige,* with Ned as their spokesman. And before Ned was through it was not merely van Tosch who was receiving the instruction; Boss and Colonel Linscomb, Lucius saw, were getting an education in their own ethic (*TR,* 281, 284, 287).

Bobo's problem, Ned explained, was that he had "got mixed up with a white man"—obviously of the inferior city sort. It was a loan shark, and Bobo, hard pressed, had only his black kinsman to turn to. "But why didn't he come to me?" van Tosch objected. "He did," Ned responded with embarrassing frankness. "You told him No." But Ned was a gentle, if exacting, tutor as he explained van Tosch's problem. "You're a white man," he reminded van Tosch. "Bobo was a nigger boy" (*TR,* 287, 288). Such, it appeared, was the fate of the plantation black who cast his lot with whites who did not understand *noblesse oblige;* rednecks got you in trouble, and, if your aristocrats did not get you out, they ought not to be surprised at how you tried to get yourself out. Boss objected that Bobo should have come to him rather than to Ned when van Tosch turned him down, but Ned cut through the sentimentality of interracial family ties. "You're a white man too," he said bluntly. Ned was the only McCaslin to whom Bobo could turn, and his response gave Boss an insight into how blacks with status felt toward shiftless types like Bobo. "Everybody," Ned pointed out, "got kinfolks that aint got no more sense than Bobo." Still, a superior black like Ned could be as tough about that as any white businessman. If his horserace scheme worked, Ned seemed to think, he had done his duty by Bobo; if it backfired, "that would a been Bobo's lookout." Ned was firm there. "It wasn't me advised him to give up Missippi cotton farming and take up Memphis frolicking and gambling for a living" (*TR,* 287, 288, 289, 293).

Now Ned was ready to lecture the impressed van Tosch on what to do with Bobo. "Keep him," he insisted; but he almost overstepped himself when he added, "Folks—boys and young men anyhow—in my people dont convince easy." Chicagoan van Tosch asked, "Why just Negroes?" Colonel Linscomb, sensing a weakness in this virtuoso performance, sardonically compounded Ned's difficulties. "Maybe he means McCaslins," he suggested. But Ned slipped this snare easily enough. The problem, he suggested, might be racial mixture: "McCaslins and niggers," he asserted, "both act like the mixtry of the other just makes it worse." But Ned wanted people to know exactly what he was talking about: "Right now I'm talking about young folks, even if this one is a nigger McCaslin." The trouble was that "maybe . . . [young folks] dont hear good. Anyhow, they got to learn for themselves that roguishness dont pay." Ned wanted van Tosch to understand how keeping an offender like Bobo—much as Lucius's father had retained Ludus—might be easier than hiring somebody else. "Maybe Bobo learnt it this time. Aint that easier for you than having to break in a new one?" (*TR*, 294).

The problems of *noblesse* were not the only ones resolved at this conclave. Faced with such a black oracle, Colonel Linscomb could not resist prodding him about racial differences. Again, Faulkner's attitudes were revealing. Six years earlier in "If I Were a Negro," Faulkner had been chiding blacks about the "virtues of responsibility" and advising them to "lift" themselves up "by . . . [their] own bootstraps." Now, in the mellow mood that helped him "understand" the "human folly" around him even when he could not "forgive it," he allowed the Colonel to overstate his case. "You [people] wake up Monday morning," Linscomb told Ned, "sick, with a hangover, filthy in a filthy jail, and lie there until some white man comes and pays your fine and takes you straight back to the cotton field . . . and puts you back to work without even giving you time to eat breakfast. And you sweat it out there, and maybe by sundown you feel you are not really going to die; and . . . [soon] it's Saturday again and you can put down the plow . . . and go back as fast as you can to that stinking jail cell on Monday morning. Why do you do it?" Ned's response came through almost like an answer to the crusading Faulkner of 1956 and allowed him a comic appeal to a "primitivism" that that earlier Faulkner seemed to have forgotten. "You can't know," Ned told the Colonel. "You're the wrong color. If you could just be a nigger one Saturday night, you wouldn't never want to be a white man again as long as you live" (*TR*, 291).

Ned's answer revealed a great deal about where Faulkner's search for a South had taken him. It had been nineteen years since Faulkner had voiced his fears about joining that "batch of decrepit old men looking stubbornly backward at a point 25 or 50 years in the past"; but he had learned what he needed to recapture the world that, with *Sartoris*, he had feared he would "lose and regret." He had learned how to separate that world from its physical location. The Mississippi Faulkner had known might be disappearing, dominated by Vardaman-style racists—soon, perhaps, to be ground under by Puritan liberals from the North. But entering the threshold of old age, Faulkner had at last learned how to recover that world: he had learned how to love it without feeling responsible for its faults. No wonder Faulkner had not been ready quite yet to "break my pencil and stop." The mellow nostalgia of his "reminiscence" had given him that "answer to the human condition" that had eluded him so long; it had given him, at last, the South for which he had been searching.

That South was a country in which the most frightening redneck was Butch Lovemaiden, in which the most frightening Yankee was a would-be aristocrat like van Tosch; and it was a country in which aristocrats and blacks understood each other as aristocrats had always been sure they did. Ned's answer to Colonel Linscomb had the ring of a final resolution of the tensions that had launched Faulkner on his search for a South. Each race had its place, Ned seemed to be saying, and there was nothing to do but live with that; but accepting one's place brought peace. For civilized aristocrats there was the honor and responsibility of *noblesse.* For primitive blacks there was the jubilee of each recurring Saturday night. There was something for everybody in Faulkner's Jim Crow Mississippi of 1905.

Notes

Abbreviations of Frequently Cited Sources

ESPL William Faulkner, *Essays, Speeches and Public Letters*, ed. James B. Meriwether (New York: Random House, 1965).

FAB Joseph Blotner, *Faulkner: A Biography*. Vols. I and II (New York: Random House, 1974).

FCF Malcolm Cowley, *The Faulkner-Cowley File: Letters and Memories, 1944-1962* (New York: Viking, 1966).

FAWP Joseph L. Fant and Robert Ashley, eds., *Faulkner at West Point* (New York: Random House, 1964).

FIU Frederick L. Gwynn and Joseph Blotner, eds., *Faulkner in the University* (Charlottesville: University of Virginia Press, 1967).

FOM Murry C. Falkner, *The Falkners of Mississippi* (Baton Rouge: Louisiana State University Press, 1967).

LIG James B. Meriwether and Michael Millgate, eds., *Lion in the Garden: Interviews with William Faulkner, 1926-1962* (New York: Random House, 1968).

MBB John Faulkner, *My Brother Bill: An Affectionate Reminiscence* (New York: Trident Press, 1963).

SLWF Joseph Blotner, ed., *Selected Letters of William Faulkner* (New York: Random House, 1977).

TLC James B. Meriwether, *The Literary Career of William Faulkner* (Princeton, N.J.: Princeton University Press, 1961).

WFO James W. Webb and A. Wigfall Green, *William Faulkner of Oxford* (Baton Rouge: Louisiana State University Press, 1965).

Notes to "Preface"

1. *FAB*, pp. 531-32.
2. *LIG*, pp. 179-80.
3. Walter J. Slatoff's *Quest for Failure: A Study of William Faulkner* (Ithaca, N.Y.: Cornell University Press, 1960) was the first major study prompted by Faulkner's statements about what he referred to as his "failure." The "ambiguity and irresolution" in Faulkner's fiction, Slatoff wrote, "at bottom, are more a matter of temperament than of deliberate artistic intent." Observing that no "final

analysis" of Faulkner's personality was presently feasible, Slatoff suggested that Faulkner's temperament was determined in considerable measure by "an almost compulsive desire to leave things unresolved and indeterminate"—a desire that led him on a "quest for failure" that insured that the tensions and ambiguities of his work would remain unresolved. Slatoff, pp. 239, 257, 259.

4. Geoffrey Hartman, "Towards Literary History," in *Beyond Formalism* (New Haven, Conn.: Yale University Press, 1970), pp. 356-58.

5. Gary Lee Stonum, *Faulkner's Career: An Internal Literary History* (Ithaca, N.Y.: Cornell University Press, 1979), pp. 25, 28. John T. Irwin in *Doubling and Incest/Repetition and Revenge: A Speculative Reading of Faulkner* (Baltimore, Md.: Johns Hopkins University Press, 1975), working from a psychiatric standpoint, suggests a somewhat similar theory through his comments on Faulkner's use of Quentin Compson in *The Sound and the Fury* and *Absalom, Absalom!*: "just as Quentin's retellings and reenactments are experienced as failures that compel him to further repetitions that will correct those failures but that are themselves experienced as failures in turn, so Faulkner's comments on his own writing express his sense of the failures of his narratives, failures that compel him to retell the story again and again" (p. 158).

6. Stonum, *Faulkner's Career*, p. 24.

Notes to "A Nice Place to Live"

1. Phil Stone, "The Man and the Land," *WFO*, pp. 5, 4.

2. Booker T. Washington, "The Atlanta Exposition Address," in Washington, *Up from Slavery* (New York: Bantam Books, 1956), pp. 157, 158.

3. Edgar Gardner Murphy, quoted in C. Vann Woodward, *The Strange Career of Jim Crow*, 2d ed. (New York: Oxford University Press, 1966), p. 108.

4. Ralph Ellison, *Shadow and Act* (New York: Random House, 1964), pp. 24, 41, 99-100.

Notes to "The Segregation of the Word"

1. William Edward Burghardt Du Bois, *The Souls of Black Folk* (New York: Signet Classics, 1969), pp. 92-93. The passage should not be read, of course, to suggest that Du Bois thought blacks could trust any group but themselves or that "the sons of the masters" were as a class reliable; his point was that blacks should recognize the complexity of the white political situation.

2. The classic discussion of Mississippi's historical political alignments is V. O. Key, Jr., "Mississippi: The Delta and the Hills," in Key, *Southern Politics* (New York: Alfred A. Knopf, 1949), pp. 229-53.

3. See Vernon L. Wharton, *The Negro in Mississippi* (New York: Harper and Row, 1965), pp. 156-80.

4. Quoted in John F. Kennedy, *Profiles in Courage* (New York: Harper and Row, 1961), p. 176.

5. Quoted in Albert D. Kirwan, *Revolt of the Rednecks* (Lexington: University of Kentucky Press, 1951), p. 18. Kirwan's book is a thorough and perceptive account of this period of Mississippi politics. See also Wilbur J. Cash, *The Mind of the South* (New York: Alfred A. Knopf, 1941).

6. Quoted in Kirwan, *Revolt of the Rednecks*, p. 134.

7. Percy, *Lanterns on the Levee* (New York: Alfred A. Knopf, 1941), p. 143.

8. Col. Falkner backed Lamar in the crucial election of 1877, which eventuated

in the Redemption. A large rally was held when Lamar spoke at Falkner's Station near Ripley; the crowd was treated to barbeque and dancing, and Falkner joined Lamar on the speaker's platform. See James B. Murphy, *L. Q. C. Lamar: Pragmatic Patriot* (Baton Rouge: Louisiana State University Press, 1973), p. 151.

9. But he had not been quite able to let politics alone, either. A sometime Whig, he had also flirted with Know-Nothingism; it was typical of the man that in an unsuccessful bid for the legislature in 1855, he had run on that ticket against John Wesley Thompson, who had adopted him after he arrived in Mississippi as a penniless teenager. The next year he served in the Electoral College, and, after the war, in addition to his backing of Lamar, he held a series of minor political offices. *FAB*, pp. 19, 39-40, 41.

10. These aspects of Falkner's story are recounted in *FAB*, pp. 9-12, 14-32, 35-36, 39, 42-45, *FAB* notes, pp. 15, 16. See also Donald P. Duclos, "Son of Sorrow: The Life, Works and Influence of Colonel William C. Falkner, 1825-1889" (Ph.D. diss., University of Michigan, 1962).

11. *FOM*, p. 7.

12. The alignment was apparently social as well. The Young Colonel acquired a home next door to the Lamars, John Faulkner wrote, and the families grew intimate. *MBB*, p. 13.

13. *FAB*, pp. 51-52, 55, 64. See also *MBB*, p. 13.

14. The Young Colonel's political maneuverings, whatever general effect they were to have on his family, may have had one very specific one. The railroad was still a thriving family enterprise, and William's father Murry, now thirty-two, wanted to run it himself, but J. W. T. Falkner, who could scarcely have failed to see the political disadvantages of owning a railroad, sold it over his son's protests less than six months after the law establishing primaries was passed. Was the sale of the railroad a political move? Faulkner's brothers had little to say in their books in explanation for the sale of the railroad. John commented tersely that his grandfather "sold it because he did not have time to attend to it" (*MBB*, p. 10). This sale left Murry bitter; he would look back on the event as having deprived him of an important chance for success. See also *FAB*, p. 68.

15. *FAB*, pp. 82, 131.

16. *FCF*, p. 67.

17. *FAB*, pp. 81, 116, 173, 244.

18. Ibid., p. 174.

19. *MBB*, p. 221.

20. *FAB*, p. 341.

21. Reed, "Four Decades of Friendship," *WFO*, pp. 181-82.

22. Quoted in William F. Holmes, *The White Chief: James Kimble Vardaman* (Baton Rouge: Louisiana State University Press, 1970), p. 193.

23. See Woodward, *Strange Career of Jim Crow*, pp. 47-59, for a discussion of the political and social aspects of this situation.

24. Percy, *Levee*, pp. 299, 306, 286.

25. Ibid., pp. 306, 273, 228, 23, 295, 309.

26. Ibid., p. 228.

27. Ibid., pp. 149, 20, 228.

28. Kirwan, *Revolt of the Rednecks*, p. 146.

29. See Holmes, *White Chief*, pp. 187-88.

30. Kirwan, *Revolt of the Rednecks*, pp. 146-47.

31. Holmes, *White Chief*, p. 197.

32. *FAB,* p. 94.
33. Dixon, *The Clansman* (New York: Doubleday, Page, and Co., 1905), pp. 50, 323.
34. *FAB,* pp. 115-16.
35. John B. Cullen and Floyd C. Watkins, *Old Times in the Faulkner Country* (Chapel Hill: University of North Carolina Press, 1961), pp. 94-96.
36. *FAB,* p. 114. Accounts of the Patton lynching are in *FAB,* pp. 113-14, and Cullen and Watkins, *Old Times,* pp. 89-93.
37. Here, and in the following pages, I have removed Faulkner's italics where appropriate.
38. Cleanth Brooks compares "Sunset" to another lynching tale, "Dry September," composed five years afterward. In the later story "Faulkner has skillfully . . . taken account of the social background of the characters," whereas "this is an element that is hardly to be found" in "Sunset" and other tales written in New Orleans during the same period. These tales were too brief for such complexity he continues, and, furthermore, "Faulkner knew that the public for whom he was writing did not want it except in the broadest terms—an obvious French Quarter rat, an obvious Italian immigrant working in a shoe-repair shop, an obvious prostitute." *William Faulkner: Toward Yoknapatawpha and Beyond* (New Haven, Conn.: Yale University Press, 1978), p. 110. This ignores the quality of the social alignment delineated in "Sunset." There is obviously no doubt about the pedigree of "Mr. Bob," and to argue that Faulkner's sailor and Cajuns are not poor whites is beside the point; to any aristocrat, these people are outsiders who do not understand the quality of the paternalistic relationship between Mr. Bob and his black dependents.

Notes to "A Visit to a Familiar Place"

1. Quoted in *FAB,* p. 415.
2. After citing a number of similarities between Charlestown, Georgia, and Jefferson, Mississippi, Cleanth Brooks argues that *Sartoris,* not *Soldiers' Pay,* should be considered Faulkner's first attempt to take Anderson's advice and write about Mississippi. "In *Soldiers' Pay* the regional setting is given no special significance" and "is not even seen in full perspective," he writes, adding that "except for the Negroes, Charlestown, Georgia, might just as well have been Charlestown, New Hampshire" (*Toward Yoknapatawpha and Beyond,* pp. 97-99). In 1966 Michael Millgate saw the book similarly: "For all the evocations of a Southern town, . . . the action remains curiously unlocalized in time or space: the local inhabitants are a composite chorus rather than individuals; the Rectory itself could be in England almost as convincingly as in Georgia; and the class-structure of the novel seems at times to be quite specifically English." *The Achievement of William Faulkner* (New York: Random House, 1966), pp. 66-67. By 1973, however, Millgate had reservations as to the novel's insularity from Mississippi: "one begins to wonder whether the choice of a Georgia rather than a Mississippi setting may not have been made purely for purposes of autobiographical camouflage." Rather, Millgate adds, "what one senses so often in *Soldiers' Pay* is the presence of autobiography minimally transferred." Millgate, "Starting Out in the Twenties: Reflections on *Soldiers' Pay,*" *Mosaic,* 7 (Fall 1973), 13, 14. Millgate presents a brief discussion of certain autobiographical aspects of the novel on these pages.

3. *FAB*, pp. 369, 410, 423, 430, 371-72. See also Anderson's fictional account of his early experiences with Faulkner in "A Meeting South" in *Sherwood Anderson Reader* (Boston: Houghton Mifflin, 1947), pp. 274-84.

4. Faulkner, David Minter writes, hoped to "find in the war that was changing the . . . [Western world] what the Old Colonel had found in the war that had changed the South: an occasion for heroism." *William Faulkner: His Life and Work* (Baltimore: Johns Hopkins University Press, 1980), p. 29.

5. For the World War I military careers of the Falkner boys, see *FOM*, pp. 88, 91, 94-103, and *FAB*, pp. 189, 207, 221-22. Judith Bryant Wittenberg asserts that "for . . . [Faulkner] as the oldest [brother] to be the only one thwarted in his effort to join the glamourous legions of the military was a psychic assault of the first order." *Faulkner: The Transfiguration of Biography* (Lincoln: University of Nebraska Press, 1979), p. 32. Of Faulkner's decision to write about the war, André Bleikasten writes: "That his motivations in choosing his subject were deeply personal is beyond doubt." Bleikasten believes that "Faulkner felt so frustrated about not having been able to take an active (and possibly heroic) part in the fight that during the postwar years one of his favorite roles . . . was the role of the wounded pilot returning from France." Bleikasten does not connect this role with the Falkners' family myth but, nonetheless, makes an interesting suggestion: Faulkner's "non-experience of the war may indeed be said to have proved as much of a shock to him as actual participation in it had to other Americans of his generation. The wound was there, even though, unlike Hemingway's, it was an imaginary one." *The Most Splendid Failure* (Bloomington: Indiana University Press, 1976), p. 17.

6. *The Letters of Sherwood Anderson*, ed. Howard Mumford Jones and Walter B. Rideout (Boston: Little, Brown and Co., 1953), p. 142.

7. Millgate writes that "no direct trace of Anderson's influence appears in *Soldiers' Pay*, although the final passage about the Negro church is faintly reminiscent of Anderson's Negroes in *Dark Laughter*, and Elizabeth Prall Anderson was later to declare: 'Nobody could influence Bill Faulkner.' " Millgate, *Achievement of Faulkner*, p. 17. Wittenberg, *Transfiguration*, pp. 38, 43, pointing out that "the similarities between the experiences of the two men sometimes make it difficult to separate 'cause' from 'accidental parallel,' " finds "resonances of Anderson in a large number of Faulkner's works." Of the use of black and white characters for contrast, she observes that "this contrast was a fundamental theme of Anderson's 1925 novel, and Faulkner may well have consciously picked it up for use in his own works." Perhaps the word "influence" gets in the way. Since both novels feature black choruses that provide commentaries on white inhibitions, some kind of intercourse seems to be going on. Faulkner, however little he might be "influenced," was capable of absorbing ideas from others, and he said so many times.

8. *MBB*, p. 183; *FAB*, pp. 52, 658.

9. "Mississippi," *ESPL*, p. 39.

10. *FAB*, p. 52.

11. *MBB*, p. 183.

12. Robert Cantwell, "The Falkners: Recollections of a Gifted Family," in *William Faulkner: Three Decades of Criticism*, ed. Frederick J. Hoffman and Olga W. Vickery (New York: Harbinger Books, 1963), p. 61.

13. *FAB*, p. 76.

14. *FOM*, pp. 13, 104.

15. "Even in his first novel," Brooks writes, "Faulkner was too good an artist to distort the ways in which his various characters would refer to Negroes." That is, Faulkner did not censor his characters; he merely reported racist attitudes where he saw them. "If Faulkner in his depiction of Negroes departs from reality, it is [not in stereotyping but] in the direction of a kind of black mystique: the Negro, it is implied, is able to see through the white man's distractions and obsessions." *Toward Yoknapatawpha and Beyond,* p. 97. Granted that one generation's well-intended epithets may turn out to be a later one's stereotypes; still, the language here quoted is generally that of Faulkner the narrator of *Soldiers' Pay,* not that of his characters, and this is only a small sampling. The reader may judge whether the implications are pejorative.

16. For an assessment of the exotic stereotype, see Nathan Irvin Huggins, *Harlem Renaissance* (New York: Oxford University Press, 1971), pp. 84-136.

17. "Negro Character as Seen by White Authors," in *Dark Symphony: Negro Literature in America,* ed. James A. Emanuel and Theodore L. Gross (New York: Free Press, 1968), p. 164.

18. *FAB,* p. 452; *FIU,* p. 58.

19. *FIU,* p. 136.

20. Thomas L. McHaney finds strong biographical elements in *Elmer* which, he feels, eventuated in the Harry-Charlotte plot of *The Wild Palms* (1939) and portions of several other works: "His continued use of the material indicates that 'Elmer' remained important for Faulkner, even though he could not bring it off in the form he had originally projected." "The Elmer Papers: Faulkner's Comic Portraits of the Artist," *Mississippi Quarterly,* 26 (Summer 1973), 282-83, 284*n.* Brooks sees the similarities between Faulkner and his hero as stemming from the fact that both are young men embarking on artistic careers and touring Europe. "But having acknowledged resemblances, . . . it is just as important to note some of the differences," he adds. "Faulkner has deliberately distanced his protagonist from himself." Elmer has "a ne'er-do-well father," his "family is continually on the move," and he "is the youngest child." In such respects Elmer does indeed differ from Faulkner—no doubt Faulkner intended to draw attention away from the analogies between himself and his hero; but there is still another possibility, and Brooks is close to it when he remarks that "Faulkner designed to have . . . [Elmer] rise, by pluck and luck, from nothing at all to social standing and wealth." *Toward Yoknapatawpha and Beyond,* p. 118. The significant analogy—at least for Elmer's youth—may well be with Colonel Falkner and his background in an itinerant family and his Horatio-Alger rise.

21. Quoted in *FAB,* p. 479. Brooks comments: "Faulkner's English Noblemen [in *Elmer*] . . . are farcical characters, arrant humors personified rather than credible human beings. Perhaps because he realized that his portraits of them were literary disasters, he soon abandoned the novel." *Toward Yoknapatawpha and Beyond,* p. 118. McHaney feels that " 'Elmer' may have been doomed from the moment the English aristocrats appear in Book III." "The Elmer Papers," p. 283. It would be interesting to know why English aristocrats gave Faulkner trouble at this stage of his career.

22. *TLC,* p. 81; *FAB,* p. 479.

23. Quoted in *FAB,* p. 538.

24. *LIG,* p. 255. Bleikasten correctly notes that "with *Sartoris* Faulkner began to realize—as some of his heroes would after him—that his own individual existence, the fate of his family, and the destiny of the South were so inextricably interwoven that, as man and writer, he had to come to grips with all of it. . . .

His novels would still be autobiography in a sense, yet to the extent only that all great literary fictions are: not mirrors of the self, nor even self-probings so much as immersions into the opaqueness of *other* selves" (*Splendid Failure*, p. 36).

25. *FAB*, p. 528.

26. For detailed analyses of Faulkner's Snopes tales see Warren Beck, *Man in Motion: Faulkner's Trilogy* (Madison: University of Wisconsin Press, 1961); Cleanth Brooks, "The Plain People," in his *William Faulkner: The Yoknapatawpha Country* (New Haven, Conn.: Yale University Press, 1963), pp. 10-28. Brooks warns (pp. 10-11) against the too easy acceptance of categories such as "poor whites" when dealing with Faulkner's South. The "well-established [literary] stereotype was a gross oversimplification," he points out, "as . . . recent scholarship has made plain." Aristocrats were a small minority, and there was great diversity among the remainder of the white population. "Faulkner is . . . aware of the niceties of a social structure which distinguishes the yeoman farmer from the tenant farmer and which sees within the category of the tenant farmer a variety of types." These are necessary distinctions in addressing the world of Faulkner's fiction as well as the world from which it was derived and should not be undervalued. They are of secondary importance, however, in assessing the question of self-identification by individuals, who historically often projected stereotypes onto groups (e.g., "hill men," "Bourbons") and then identified themselves in opposition to such stereotypes. In such a scale, those in the middle—often enough, yeoman farmers who were no more than a generation away from dirt farms and poverty—were likely to identify themselves along the lines of their ancestors or of their fundamentalist religion; others—the Falkners themselves during the time of the Old Colonel—were likely to identify themselves as the aristocrats they hoped to become. See also James Gray Watson, *The Snopes Dilemma: Faulkner's Trilogy* (Coral Gables, Fla.: University of Miami Press, 1968).

27. *FAB*, pp. 531-32.

28. Ibid., pp. 531-32.

29. Ibid., p. 532.

30. *FIU*, p. 285.

31. When *Flags in the Dust* was cut to produce the manuscript published as *Sartoris*, the phrase was dropped—although it was used to describe other characters. Since all of the cutting was done by Ben Wasson—Faulkner refused to have anything to do with it (*FAB*, pp. 582-84)—it may be taken as his intention so to describe Byron.

32. *FIU*, p. 254.

33. Compare *FAB*, p. 35.

34. John Faulkner was sure that Auntee was the prototype of the "undefeated spinster aunts": "She was the Miss Jenny, the Granny Millard, all the women in *The Unvanquished* that Bill wrote about" (*MBB*, p. 70).

35. Wittenberg (*Transfiguration*, p. 70) believes old Bayard to be so "closely modeled on Faulkner's grandfather" that "there are almost no points here where fact and fiction diverge."

36. *FIU*, pp. 23, 250. Wittenberg believes that the Bayard-John tension grew out of Faulkner's own feelings toward his younger brother Dean, later killed in Faulkner's plane. "In his rejection of his family, his pessimism, his self-destructive and regressive impulses, and in the powerful guilt that arises from a subconscious need to obliterate his brother, . . . [young Bayard] is an exaggerated psychological self-portrait of the author himself." *Transfiguration*, p. 69.

37. Faulkner had leaned heavily on family servants like Ned Barnett and Caro-

line Barr in *Soldiers' Pay*, and for Simon he drew on Ned again. He took the name "Strother" from another family who served the Falkners. Joby Strother was a youth who worked around Murry Falkner's house and stable and sometimes played with Faulkner and his brothers (*FAB*, pp. 87, 538). Faulkner had had Joby on his mind (Ibid., p. 47) when he first conceived *Flags in the Dust*, and he gave his name to Simon's grandfather, John Sartoris's servant.

Notes to "A Visit with an Idiot"

1. *FIU*, p. 251. Wittenberg (*Transformation*, p. 71) comments: "Although Faulkner gives no satisfactory reason for completely glossing over the fictional generation to which his own father belonged, it is conceivable that he did so either because he was as yet unprepared to deal artistically with Murry or because he envisaged omission as some sort of punishment."

2. *LIG*, p. 222.

3. *FAB*, p. 105. Concerning the origins of the novel in Faulkner's consciousness, Bleikasten quotes André Green to the effect that the act of writing "presupposes a wound, a loss, a bereavement, which the written work will transform to the point of producing its own fictitious positivity. . . . Reading and writing are a ceaseless work of mourning. If there is a pleasure to be found in the text, we always know that this pleasure is a surrogate for a lost gratification, which we are trying to recover through other means." Green, "Le double et l'absent," *Critique*, 29 (May 1973), 403-4. Thus for Bleikasten, Caddy, " 'the beautiful and tragic little girl' [in Faulkner's words] whom he set out to create through the power of words was manifestly intended to fill a vacancy"—the daughter he had wanted but had thus far been denied. Noting also that "the seminal image of the novel is focused on the grandmother's death," Bleikasten proceeds: "Mourning, then, is not only a possible key to the process of Faulkner's creation, but a motif readily traced in . . . [his] novels themselves. . . . One would like to know . . . what its emergence at this point [at the time of *The Sound and the Fury*] means in psycho-biographical terms; yet, apart from the hints one can find in Faulkner's comments and above all in his fiction, there is unfortunately little to gratify our curiosity." Bleikasten, *Splendid Failure*, pp. 52-53. Since Faulkner's sense of loss was projected onto two characters (Caddy and Damuddy), it is natural to suggest that that loss was biographically focused on one or more persons; but why not, also, on what those persons represented to Faulkner? This is particularly important since the sense of loss that prompted *Sartoris* was ascribed by Faulkner not to a particular person but to a "world" of things he feared he would "lose and regret."

4. *LIG*, pp. 146, 222. As Faulkner reminisced about the origins of his novel, he made an interesting slip. There was only one girl in that group in *The Sound and the Fury*; but if "one of the little girls" climbed the tree, there were at least two in his original conception of the novel, and there were two in the group that played with the Falkner boys around the time of Granny Falkner's death. One was six-year-old Sallie Murry Wilkins. "Throughout our childhood," Jack Falkner wrote, "Sallie Murry and Bill and John and I could not have been any closer had we been sister and brothers" (*FOM*, p. 7). It was, after all, Sallie Murry's grandmother who had died; but the other girl close to the brothers was ten-year-old Estelle Oldham, who married Faulkner in 1929, and Estelle was big enough to climb the tree behind the Big Place that the children (*FAB*, p. 566) were fond of climbing. Whatever Caddy's origins, she brought together something of Faulkner's childhood memories of both girls. See *FAB*, p. 568.

5. *LIG,* pp. 146-47. The sequence in which Faulkner conceived all this varied slightly in his accounts. See ibid., pp. 222, 244-45; *FIU,* pp. 1, 17, 31-32, 63-64, 84; *FAB,* pp. 566-71, 577.

6. Years later Faulkner hinted enigmatically of himself and Quentin: "Ishmael is the witness in *Moby Dick* as I am Quentin in *The Sound and the Fury." FAB,* p. 1522. As Quentin appears in Faulkner's short stories and novels, concludes Estella Shoenberg, he "is but a thinly disguised autobiographical figure, and his dilemma of whether to love or to hate, to credit or to denounce, is the dilemma of his creator." *Old Tales and Talking* (Binghamton, N.Y.: University Press of Mississippi, 1977), p. 149; see also pp. 7-15.

7. As Jack Falkner reminisced about his father, he allowed himself a measure of irony. Of the family legend that the sale of the railroad had frustrated Murry Falkner's hopes of an outstanding career, Jack commented that his father "could certainly always reflect that few men become vice-president of a railroad at twenty-five." Mr. Murry's "good qualities were legion," he insisted, "and they . . . did much to mold the characters of his sons"; but he could not resist adding that his father "was not an easy man to know." Mr.Murry's "capacity for affection was limited," he thought, and the best he could say of that was that "to such an extent as it allowed he loved us all" (*FOM,* pp. 11-12). Minter writes that "Murry Falkner became widely regarded not only as a failure but also as a drinker" and that "few of his jobs held any intrinsic interest for him." In his declining years he spent "much of his time . . . alone, sitting in silence." Faulkner "remained outwardly respectful [to his father], but he thought of him as an embarrassing failure and a dull man." Minter, *Life and Work,* pp. 15, 8-9, 16. Of the relationship between Mr. Murry and Mr. Compson, Wittenberg writes: "Mr. Compson is like Faulkner's father only in his alcoholism and detachment, for he is cerebral and articulate, a lawyer who knows Latin. . . . But though Mr. Compson shows moments of compassionate understanding, he is as destructive a male parent as was Murry Falkner, chiefly because his only legacy to his son is a constant stream of nihilistic statements." Wittenberg believes that in some ways Quentin's brother Jason reflects Mr. Murry's personality. "Faulkner revealingly gave Jason some of his father's characteristics, his verbal idiom, his temper, his lifelong sense of having been deprived of a job that he wanted badly, and a gruffiness that is as lacerating as Faulkner undoubtedly once perceived his father's to be." *Transfiguration,* pp. 81-82, 84-85.

8. Miss Maud, who came from a very different family background than her husband's, was the achiever of the household. Her family, people said, was one of those that had never recovered from Reconstruction, but there were other problems, too. Charles Butler had been responsible enough to serve as sheriff of Lafayette County; but he had run into debt, and, after his creditors closed on him in 1888, he simply disappeared—rumor had it that he had run off with a beautiful octoroon. Whatever the enticement, he had left his wife and children in pitiful financial condition, and that poverty had shaped their lives until Miss Maud married Mr. Murry (*FAB,* pp. 57-58). Minter writes that "in the mother who made him perfectly aware of his father's weakness and then forced him to choose between that weakness and her strength, . . . [Faulkner] saw fierceness that went too far," adding that Faulkner suffered from a "continuing dependence on his mother." Minter, *Life and Work,* pp. 17, 19. As Wittenberg (*Transfiguration,* p. 81) points out, "the one psychoanalyst to whom Faulkner ever consented to talk concluded that the author felt his mother had given him all too little emotional sustenance." (For this incident, see *FAB,* p. 1454.) Wittenberg conjectures

that "in his portrayal of Mrs. Compson . . . Faulkner may have been drawing upon both his sense of his mother's affectional failings and his memories of a specific period in 1915 when he was Quentin's age and his mother was ill for several months and he and his brothers had to fend for themselves." Compare *FAB,* p. 178.

9. In quoting from *The Sound and the Fury,* and from other Faulkner novels, I have omitted Faulkner's italics when they have no direct bearing on my point.

10. Faulkner drew up a chronology of the events in his novel. See *FAB,* p. 572. Faulkner never specified whether Damuddy was a Compson or a Bascomb. Since Mr. Compson remained composed at her death and Mrs. Compson wept, one assumes Damuddy was Mrs. Compson's mother.

11. Or, as Brooks puts it, "Quentin is emotionally committed to the code of honor, but for him the code has lost its connection with reality; it is abstract, even 'literary.' Quentin's suicide results from the fact that he can neither repudiate nor fulfill the claims of the code." *Yoknapatawpha Country,* p. 337.

12. Bleikasten writes, "Quentin's is the Puritan version of the Fall: Eve was the beginning of evil; it was through her that the innocence of Eden was lost. Although he clings to the Southern myth of Sacred Womanhood, and will defend Woman's honor to the last, . . . Quentin's faith in her purity has been shattered by his sister's sexual misconduct." *Splendid Failure,* p. 99. Brooks identifies Quentin as "another of Faulkner's Puritans" and, without conjecturing why this son of an aristocratic family should be a Puritan, asserts that he "reveals his Puritanism most obviously in his alarm at the breakdown of sexual morality. When the standards of sexual morals are challenged, a common reaction and one quite natural to Puritanism is to try to define some point beyond which surely no one would venture to transgress—to find at least one act so horrible that everyone would be repelled by it." *Yoknapatawpha Country,* pp. 331-32. Brooks cites Faulkner's Compson genealogy, published almost a decade after *The Sound and the Fury,* as proof: Quentin, Faulkner said then, "loved not the idea of the incest which he would not commit, but some Presbyterian concept of its eternal punishment." "Appendix: The Compsons," in *The Portable Faulkner,* rev. ed. (New York: Viking, 1967), p. 710. To say that Quentin adopted "*some* Presbyterian concept" [emphasis added] is obviously not the same as saying Quentin was "another of Faulkner's Puritans"; quite possibly Faulkner meant that Quentin adopted that concept for the sake of having a concept of sin on which to base his fantasy of a hell with Caddy. It is another thing, however, to describe this [in Faulkner's genealogy] descendant of the heroes of Culloden as a Puritan experiencing a merely Puritan shock at his sister's sexuality. Quentin struggles to associate his Cavalier image with the supposedly highborn Compson males, and he wants to adopt an image of Caddy as a Cavalier *lady*; but his image of women is inherited from his mother's supposedly less highborn Bascomb blood and his dilemma is that Caddy has behaved in a fashion unbecoming to a lady. This quandary parallels his fear, dramatized in *Absalom, Absalom!,* that the Puritan rather than the Cavalier myth represents the "real" truth about the South. Quentin wants to be a Cavalier but is sometimes afraid that he is being forced by circumstances to adopt a Puritan vision.

13. Quentin inherited his failure, Faulkner said later. "The action as portrayed by Quentin was transmitted to him through his father. There was a basic failure before that. The Grandfather had been a failed brigadier twice in the Civil War. It was the—the basic failure inherited through his father, or beyond his father" (*FIU,*

p. 3). This, Bleikasten points out, is "a perspective barely outlined in the novel itself," yet nonetheless there: "Beyond his individual failure lies the bankruptcy of at least two Compson generations (Appendix, 408-9), and beyond that again the disease and decay of a whole culture." But Quentin's allegiance to that culture is not a matter of choice, Bleikasten points out. "Through his father, he is heir to the Southern tradition, to its code of honor. . . . When this pattern of values is passed on, however, it has already lost its authority, the more so in this case as the appointed transmitter of the Southern creed [Mr. Compson] is an inveterate skeptic. Quentin clings to it because to him it is the only available recourse against absurdity, and because its very rigidity seems a safeguard of order and integrity." Still, his "fidelity is an allegiance to values long dead." *Splendid Failure,* pp. 110-11.

14. *LIG,* pp. 147, 245. For a study of the aesthetics implied in this statement, see Margaret Blanchard, "The Rhetoric of Communion: Voice in *The Sound and the Fury,*" *American Literature,* 41 (Jan. 1970), 555-65.

15. Faulkner's interest in primitivist painting may have come from a closer source: Miss Maud was an accomplished primitivist painter.

16. *The Autobiography of an Ex-Colored Man* (New York: Alfred A. Knopf, 1970), pp. 173, 175. Bruce A. Rosenberg has written of the significance of the traditions of the black sermon for Faulkner's novel in "The Oral Quality of Reverend Shegog's Sermon in William Faulkner's *The Sound and the Fury,*" *Literature in Wissenschaft und Unterricht,* 2 (1969), 73-88.

17. Compare the Reverend C. C. Lovelace, "The Wounds of Jesus," as transcribed by Zora Neale Hurston at Eau Gallie in Florida, May 3, 1929. In 1924 Hurston used the transcription in her novel, *Jonah's Gourd Vine* (Philadelphia: Lippincott, 1971), pp. 269-81.

18. Hailing the Shegog sermon as "a triumph of Faulkner's verbal virtuosity," Bleikasten notes that "this triumph was achieved through a gesture of humility. For the novelist scrupulously refrained from improving on the tradition of the oral sermon as he found it. There is no literary embellishment." *Splendid Failure,* p. 200.

19. Millgate argues that "Faulkner does not intend any simple moral division between the Negroes and their white employers," using Luster as an illustration of a black who is not touched by the Easter sermon. Faulkner is too complex in his rendering of the problem for any such absolute distinction as Millgate speaks of; however, no white person in the novel is revealed as capable of understanding the Easter sermon, and the majority of the black congregation do participate in it. Brooks is closer to the mark: "Faulkner's Negro characters [in *The Sound and the Fury*] show less false pride, less false idealism, and more seasoned discipline in human relationships [than whites]. Dilsey's race has also had something to do with keeping her close to a world still informed by religion. . . . The Compson family—whatever may be true of the white community at large . . . —has lost its religion." See Millgate, *Achievement of Faulkner,* p. 102; Brooks, *Yoknapatawpha Country,* p. 344.

20. When Faulkner revisited his characters from *The Sound and the Fury* for *The Portable Faulkner* in 1946, he wrote for the section devoted to Dilsey only two words: "They endured." *The Portable Faulkner,* p. 756. This statement, which connects Dilsey with the theories of black endurance enunciated by Isaac McCaslin in *Go Down, Moses* (1942), has attracted much attention. Millgate rightly notes the qualified nature of Dilsey's endurance: "Dilsey 'endures,' but her

endurance is tested not in acts of spectacular heroism but in her submission to the . . . demands made upon her by the Compson family." *Achievement of Faulkner,* p. 101. The Faulkner of *The Sound and the Fury* shunned pronouncements of the type he later ascribed to Isaac; amid the concreteness with which Dilsey was realized in the earlier novel, they would have been out of place. The Faulkner who theorized about black virtues like "endurance" was a later development.

21. Although he believes passages in *Sartoris* imply "condescending mimicry," Irving Howe writes that "none other" among American novelists "has listened with such fidelity to the nuances of . . . [black] speech and recorded them with such skill." *William Faulkner: A Critical Study* (New York: Vintage Books, 1952), pp. 121, 134.

22. When Malcolm Cowley was putting together *The Portable Faulkner,* Faulkner suggested that he include "for the sake of the negroes, that woman Dilsey who 'does the best I kin.' " *FCF,* p. 25.

23. "I worked so hard on that book," Faulkner said later, "that I doubt if there's anything in it that didn't belong there" (*FAB,* pp. 589-90). But he never made his intentions clear about either Roskus's death or the status of Frony's husband (Luster's father). In the chronology he made for his own use as he constructed the novel, Faulkner noted that Roskus died in 1915 (*FAB,* p. 572), but he did not pin that event down precisely in his text. On one occasion in Benjy's narrative, the moaning at the Gibson home seems to be for someone in their black family: "*They moaned at Dilsey's house. Dilsey was moaning. When Dilsey moaned Luster said, Hush, and we hushed, . . . and Blue howled under the kitchen steps* (*SF,* 24). The passage is italicized, and when the narrative is italicized again a few lines later Benjy may be thinking of the same incident: "*Dilsey moaned, . . . and Blue howled under the steps. Luster, Frony said in the window, Take them down to the barn. I cant get no cooking done with all that racket*" (*SF,* 25). If the two passages are about the same incident, the mourning would seem to be for Roskus, not for Luster's father, because Luster and Frony (who has taken over the household temporarily) are the composed ones in the group. If the Gibsons are grieving for Roskus's death, Luster, in the second passage, must be referring to his ghost: "*I aint going down there, Luster said. I might meet pappy down there. I seen him last night, waving his arms in the barn*" (*SF,* 25). In revisions of his manuscript, Faulkner sometimes changed speeches attributed to T. P. to Luster (*FAB* Notes, p. 87), and he may have done so here; but there is no reason Luster, who calls Dilsey "mammy," should not call his grandfather "pappy." Luster's father seems to be alive on April 7, 1928, when Dilsey threatens Luster, "You just wait till your pappy come home" (*SF,* 45). But if he is alive, he is obviously not in Jefferson. Whatever Faulkner's intentions, he failed to clarify either Roskus's death or the absence of Frony's husband, and, when he put together his history of the Compsons for Cowley, he was not much more helpful. He reported that Frony was now married to a Pullman porter and had gone to live in St. Louis, then had come back to Memphis to make a home for the aging Dilsey; but he never connected Frony's porter to Luster, and he never mentioned Roskus.

24. Asserting that the best in Dilsey's characterization resulted from the ability of a "gifted" artist to "salvage significant images of life from the most familiar notions," Howe points out that despite her "hard realism" and "her ability to maintain her selfhood under humiliating conditions," still "the conception behind Dilsey does not seriously clash with the view of the Negro that could be held by a white man vaguely committed to a benevolent racial superiority." He adds: "I

should like to register a dissent from the effort of certain critics to apotheosize her as the embodiment of Christian resignation and endurance. The terms in which Dilsey is conceived are thoroughly historical, and by their nature become increasingly unavailable to us: a fact which, if it does not lessen our admiration for her as a figure in a novel, does limit our capacity to regard her as a moral archetype or model." *Critical Study,* p. 123. Lee Jenkins asserts that Faulkner "misrepresents the meaning and significance of black life, in relation to that of the whites, and diminishes and demeans blacks in comparison. This is the case even though the blacks, in the person of Dilsey, appear to exemplify a dignity and an endurance lacking in the whites. It is the dignity and endurance, according to the terms of her creation, of a victim who conspires in her own victimization." *Faulkner and Black-White Relations* (New York: Columbia University Press, 1981), p. 163. Bleikasten takes a somewhat different approach. Although Dilsey "seems to fit rather nicely in the tradition of the black mammy," he asserts that "the point is that her virtues, as they are presented in the novel, owe nothing to race." Which, of course, they do not; but they owe a very great deal to the literature through which Faulkner's class attempted to justify its racial attitudes—as Bleikasten implies when he notes that Dilsey's "literary lineage is readily traced back to Thomas Nelson Page." *Splendid Failure,* p. 191. See also George E. Kent, "The Black Woman in Faulkner's Works, with the Exclusion of Dilsey," Parts I and II, *Phylon,* 35 (Dec. 1974), 430-41; 36 (Mar. 1975), 55-56.

Notes to "A Visit with Some Puritans"

1. Percy, *Levee,* p. 226.

2. Ellison, *Shadow and Act,* pp. 47-58.

3. Brooks works out an elaborate analysis of Faulkner's use of the traditions of courtly love (with their Manichaean implications) in *The Town. Yoknapatawpha Country,* p. 197ff. Regrettably he does not extend this analysis to *Light in August.* In the following pages I have sketched no more than the bare outlines of the implications of Manichaeanism for this and other of Faulkner's novels.

4. *FIU,* pp. 173, 189-90.

5. Faulkner's "Negro murderer" again revealed his closeness to writers who had established the exotic image. "I stole that character from Roark Bradford," he told Marshall Smith in 1931. "He told me that story one night while we were just talking. I waited two years for him to use it, but he never did" (*LIG,* p. 11). This black's dialect, which sounded more Louisianian than Mississippian ("Ise a gawn po sonnen bitch" [*S,* 92]), reinforced Faulkner's statement. John Cullen had a different story. The "Negro murderer" was a local black, Cullen claimed, named Dave Bowdry: "He was the Negro who was held in the murderer's cell behind the heaven tree which still stands near the [Oxford] jail, and he sang songs as Faulkner says in *Sanctuary*" (Cullen and Watkins, *Old Times,* p. 73).

6. "Dry September" appears to have been first conceived in 1930. See *FAB,* pp. 646-47.

7. Brooks writes that "one feature of 'Dry September' that sets it apart from most stories about lynching violence is the way in which Faulkner has . . . taken account of the social backgrounds of the characters. . . . Faulkner does not explicitly refer to the social background of each of his characters, but allows the background to emerge through the speech and act and gesture of the character. He has, for example, been careful to place 'Miss Minnie' in exact relation to the

community as we learn about her as an individual." For example, Brooks quotes from the story: "She was of comfortable people—not the best in Jefferson, but good people enough" (*CS*, 173-74). *Toward Yoknapatawpha and Beyond*, p. 110.

8. Originally titled "Never Done No Weeping When You Wanted to Laugh," this story evidently was conceived prior to *The Sound and the Fury*. Faulkner almost doubled its length before it appeared in *American Mercury* in 1931 as "That Evening Sun Go Down"; see *FAB*, pp. 565-66, 688; *FAB* Notes, p. 82. Cullen believed this story was cut from the same cloth as Faulkner's tale of a "Negro murderer" in *Sanctuary*. "Dave [Bowdry] was a probable source for . . . the husband of Nancy in 'That Evening Sun.' . . . Dave did knock his wife in the head, cut her throat as though butchering a hog, and throw her body behind the bed. . . . There is a ditch like the one Nancy had to cross behind the place where the Faulkners used to live. Dave committed the murder a short distance from the Faulkner home." See Cullen and Watkins, *Old Times*, pp. 72-73.

9. Nancy—in Faulkner's imagination at least—would survive her fears to receive the Word. Twenty-two years later, after he had brought her back as a redeemed Christian in *Requiem for a Nun* (1951), he insisted she was "the same person" as in the earlier work. *FIU*, p. 9.

10. *FIU*, p. 199.

11. Ibid., p. 97.

12. *FAB*, pp. 133, 703.

13. Joe Christmas's tragedy evolved slowly in Faulkner's mind. In an early version of the manuscript he had what Faulkner then referred to as "black blood" and knew it. See Regina K. Fadiman, *Faulkner's* Light in August: *A Description and Interpretation of the Revisions* (Charlottesville: University of Virginia Press, 1975), pp. 42-43.

14. For a commentary on the curse of Cain, see Alan W. Watts, *The Two Hands of God: The Myths of Polarity* (Toronto: Collier Books, 1969), pp. 121-26. Jenkins (*Black-White Relations*, pp. 58-60) identifies the curse as the curse of Ham.

15. That theology was curious for a Puritan, much less an abolitionist. But Faulkner was taking pains to suggest that the Burdens' faith (and by inference, that of his other Puritans) was more pagan than Christian. The measure of Calvin's paganism was his illiteracy in English. It was because he could not spell his father's name, Burrington, that Nathaniel's father became Calvin Burden, bearer of the black cross. Taught Spanish by California monks, Calvin read to his children out of a Spanish Bible. He thought the Spanish tainted its message, and he tried to embellish its "fine, sonorous flowing of mysticism in a foreign tongue" with his own "harsh, extemporized dissertations." Calvin had to follow the oral rather than the written traditions of his people, and his memory neatly divided his cosmos into light and darkness; his lectures to his family were "composed half of the bleak and bloodless logic which he remembered from his father on interminable New England Sundays, and half of immediate hellfire and tangible brimstone of which any country Methodist circuit rider would have been proud" (*LIA*, 179). The Burdens, Faulkner seemed to be showing, had inherited the spirit more than the doctrine of Puritan teaching, and the spirit was pagan. Calvin was buried in a grove, a sacred place of Nordic paganism. When Nathaniel had told Joanna about the curse of Cain, he had taken her there, not to a church. And Joanna was sure that curse had something to do with the grove. It was "a something that I felt that . . . [Nathaniel] had put on the cedar grove," she remembered, "and that when I went into it, the grove would put on me" (*LIA*, 187).

Notes to "How to Visit the Black South"

1. *Black Boy: A Record of Childhood and Youth* (New York: Harper and Brothers, 1945), p. 228.
2. *Native Son* (New York: Perennial Classics, 1966 [orig. publ. 1940]), p. 108.
3. Baldwin, "Many Thousands Gone," in *Notes of a Native Son* (Boston: Beacon, 1955), pp. 37-38, 41.
4. "How 'Bigger' Was Born," in Wright, *Native Son*, p. xxxiv.
5. Wright, Baldwin, and Ralph Ellison were all familiar with Faulkner's novels. Surprisingly, none of the three displayed interest in *Light in August.*
6. Ellison, *Shadow and Act*, p. 304.
7. *FIU*, p. 72; see also pp. 97, 118.
8. Ellison, *Invisible Man* (New York: Random House, 1952), pp. 3, 4.
9. Baldwin, "Many Thousands Gone," p. 38.
10. *FIU*, pp. 118, 72.
11. Cleaver, *Soul on Ice* (New York: Delta Books, 1968), p. 14.
12. Ibid., p. 14.
13. Ibid., pp. 8, 10-11.
14. Ibid., p. 159.
15. Wright, *Native Son*, pp. 357, 360, 364, 366, 354.
16. *FIU*, p. 72.
17. Baldwin, "Many Thousands Gone," p. 44.
18. Brooks argues that Joe's death at Grimm's hands should not be called a "lynching": "A lynching is defined as the concerted action by private individuals who execute summary punishment outside the forms of the law." Insisting that "something more than a mere quibble about a term is involved," he argues that "if we use the word 'lynching' loosely and carelessly, we shall be in danger of missing the relation of Joe Christmas to the community he has defied, and . . . of Percy Grimm to the community he claims to represent. There is, in fact, every reason to think that Grimm's whole *conscious* motivation is to insure that the good name of the town not be marred by a lynching—even though the community itself is not apprehensive of a lynching." *Yoknapatawpha Country*, pp. 51-52. This reasoning obscures Faulkner's intentions. Are we to believe, then, that lynchings are a wholly *conscious* activity of a community? Or that the sheriff and the local American Legion commander, whom Brooks cites, believe that no lynching may occur? Or that none of Grimm's followers, who are also chasing Joe, would have killed him as Grimm does? Brooks's arguments draw attention away from the inevitability Faulkner portrays in the ritual of scapegoating that is being acted out between the community and Joe and hence distort the reader's apprehension of the central aims of the novel. Brooks is closer to the mark in his statement about Will Mayes of "Dry September"—a statement that is also a fair assessment of what happens to Christmas: "He owes his death in great part to a social climate." *Yoknapatawpha and Beyond*, p. 109.
19. Wright, *Native Son*, p. 354.
20. *FIU*, p. 72.
21. Baldwin, "Many Thousands Gone," p. 44.
22. Ibid., pp. 44, 35, 34.
23. Ibid., p. 35.
24. Ibid., pp. 44-45, 40.
25. Ibid., p. 44.

Notes to "A Visit to a Dark House"

1. For the history of Rowan Oak, Judith's ghost, and Faulkner's efforts to make the place liveable, see *FAB*, pp. 651-53, 657-61. Wittenberg (*Transfiguration*, p. 119) comments that Faulkner now had found for himself a "newly masculine role": "Within barely a year, . . . [he] had become husband, stepfather, homeowner, renovator, and chief-of-staff to a tiny army of servants, and the final affirmation of . . . [this] role came when he made his wife pregnant during the summer." He had, of course, gone from the role of itinerant artist to that of landowner/father/husband that resembled the hierarchic roles of Falkner males. Minter (*Life and Work*, p. 122) writes that Faulkner "wanted to establish a home so clearly evocative of his family's past that it would make him the acknowledged center of his clan." Faulkner's fascination with these roles now became an even stronger preoccupation of his fiction. Shoenberg (*Old Tales and Talking*, pp. 65-66), emphasizing the hold on Faulkner's imagination claimed by the Shegog legends, believes that Shegog "was—if anyone was—the historical figure on whom Colonel Thomas Sutpen was based." Shoenberg is particularly interested in the role played by Judith Sutpen—based, perhaps, on the legends of Judith Shegog—in "Revolt of the Earth," an abortive film script written sometime after *Absalom, Absalom!* in collaboration with screenwriter Dudley Murphy. "Like Judith Shegog . . . , whose ghost was supposed to haunt the house Faulkner lived in . . . , Judith Sutpen is woed by a Yankee soldier; but unlike Judith Shegog, who broke her neck trying to descend a ladder in an elopement attempt, this Judith lives to marry her Yankee." Later, she returns to the plantation, where, instead of haunting it, she is haunted by Sutpen's other descendants, black and white.

2. *FAB*, pp. 701-2.

3. William Faulkner, *Uncollected Stories*, ed. Joseph Blotner (New York: Random House, 1979), pp. 606, 604. Although it was in the summer of 1931 that Faulkner seems to have committed himself in earnest to "Evangeline," a previous version—or versions—probably existed. See *FAB*, p. 696.

4. Faulkner, *Uncollected Stories*, pp. 608, 609.

5. Ibid., p. 600. Blotner calls attention to this passage, noting that "the story would not stay quiet in his mind" and that he "would not allow the tale to remain pointless." *FAB*, pp. 698, 699. For a discussion of the composition of "Evangeline" and its relation to the rest of Faulkner's work, see Shoenberg, *Old Tales and Talking*, pp. 30-49.

6. Shoenberg (*Old Tales and Talking*, pp. 59, 60, 63) writes that "it is hard to tell whether 'Wash' was written before or after 'Evangeline' . . . or even before the main part of the novel." "Wash" apparently originated as one of Faulkner's Snopes tales and was conceived rather early in his career but written in its final form at a later date, but yet subsequent to the composition of "Evangeline." Shoenberg concludes: "For *Absalom, Absalom!*, it seems, Faulkner took the relatively unimportant figure of Colonel Sutpen in 'Evangeline' to tell in 'Wash' how and why at the last possible moment, when Sutpen's crumbling design could have been restored by the birth of a son, Wash Jones killed Sutpen." The matter is complicated by the discovery of a fragment—which Shoenberg says "was typed in 1925 or earlier"—containing a dialgoue between characters named "Clytie" and "Wash." Shoenberg quotes McHaney's statement ("The Elmer Papers," p. 284*n*8) that "there is startling evidence in Faulkner's early work that most of his characters and thematic preoccupations existed in his mind almost from the beginning of his career." This,

Shoenberg says, "points up the extreme difficulty of tracing any thematic or substantive thread through Faulkner's work, which is not—McHaney's comment is well-founded—a sequence but a galaxy of continuums." Suggestive as these considerations are for Faulkner's period of gestation in the mid-1920s, they can be misleading when applied to his career and its continuity. If followed to their conclusion, they threaten to rob Faulkner of any credit for intellectual or aesthetic development. To have conceived the plots of most of his work during that gestation period is one thing; but not to have undergone any significant development in the years that followed, or merely to have pieced these tales together in new combinations, is another. The dynamic element in Faulkner's career is not his origination of these tales, but his mode of perceiving them.

7. To Harrison Smith, *SLWF,* p. 79.

8. Shoenberg has inventoried the stories in which Quentin is or may have originally been the narrator (*Old Tales and Talking,* pp. 16-29). She identifies two Quentins, one who is a suicide, the other "the born and bred story-teller" (p. 29). Shoenberg stops short of identifying a dynamic in the development of these personae, and to suggest their meaning for Faulkner, employs the figure used by Quentin in *Absalom, Absalom!*—of a pebble dropped in water and producing concentric ripples. "It is almost fortunate for this discussion that most of Faulkner's short stories have not been reliably dated. If their dates of composition were known it would be easy to forget that Faulkner, like Quentin, had 'always known' these stories; and it would be tempting to 'organize' them into a chronology, a straight line, which would destroy the ripple figure Quentin himself provided" (p. 28). What remains, however, is the fact that on certain dates Faulkner used Quentin, on others he did not, and on still others he used other characters who resembled Quentin with slight modifications. What is important is his motivation in making such changes and for making them in a particular sequence.

9. *SLWF,* p. 79.

10. *FIU,* pp. 281, 76, 36.

11. Blotner dates Faulkner's letter to Smith outlining his purposes with *A Dark House* as "probably Feb. 1934" (*SLWF,* p. 78). Possibly in mid-April 1934 he sent agent Morton Goldman an aviation story, "This Kind of Courage" (see *FAB,* p. 841), from which he later evolved *Pylon.* Blotner dates Faulkner's letter to Goldman, which accompanied "Ambuscade," as "probably late Spring 1934" (*SLWF,* p. 80). Meanwhile, he had continued to work on *A Dark House* (see *FAB,* pp. 832, 841). Faulkner's letter to Goldman, which accompanied "Retreat," is dated by Blotner as "probably late Spring 1934" (*SLWF,* p. 80), and "Raid" followed shortly (*FAB,* pp. 848-49). By August he was writing Smith that "the only definite news" he had about the novel was "that I still do not know when it will be ready," as it was "not quite ripe yet." Although he had "to put it aside and make a nickel every so often," he thought "there must be more [to his inability to finish] than that." He had "a mass of stuff, but only one chapter that suits me" (*SLWF,* pp. 83-84). In September he sent "Riposte in Tertio" (then titled "The Unvanquished") and "Vendée" to the *Post* (*FAB,* p. 858), and a story called "Drusilla"—later retitled "Skirmish at Sartoris"—in October (*SLWF,* p. 85). By October he was writing Goldman that he wanted "This Kind of Courage" back, as he was "writing a novel out of it" (*SLWF,* p. 85). The publication dates of the series were: "Ambuscade," *Saturday Evening Post,* 207 (Sept. 29, 1934); "Retreat," ibid., 207 (Oct. 13, 1934); "Raid," ibid., 207 (Nov. 3, 1934); "The Unvanquished" ("Riposte in Tertio"), ibid., 209 (Nov. 14, 1936); "Vendée," ibid., 209 (Dec. 5, 1936); and

"Skirmish at Sartoris," *Scribner's Magazine,* 97 (Apr. 1935). Citations hereafter are from these sources.

12. *FIU,* p. 252.

13. Minter (*Life and Work,* p. 144) writes that "in creating Old Bayard's youth, . . . [Faulkner] drew on several family stories, including some the Young Colonel had told him."

14. "Retreat," p. 89.

15. "The Unvanquished," p. 128.

16. "Vendée," pp. 16, 94.

17. "Skirmish at Sartoris," pp. 199, 200.

18. When Faulkner brought the series together as *The Unvanquished* (New York: Random House, 1938), he added passages explaining these relationships. See pp. 14, 16.

19. "Ambuscade," p. 80, and "Retreat," p. 87.

20. "Raid," p. 73.

21. In "Mississippi" (1954), Faulkner reported that Ned Barnett was "born in a cabin in the back yard in 1865, in the time of . . . [Faulkner's] great-grandfather." *ESPL,* p. 39. Ned, who died in the winter of 1947 (*FAB,* p. 1243), had come to Oxford to work for J. W. T. Falkner "after the death of the [Old] Colonel" (*FAB,* p. 52). According to John Faulkner, Ned had "been a slave belonging to my great grandfather" (*MBB,* p. 183); according to Blotner, he continued to think of himself in that fashion (*FAB,* p. 52). If William Faulkner was correct, Ned's having been "a slave" would seem to be a technicality; at all events, that would make Ned at least eighty-two when he died, seventeen years younger than J. W. T. Falkner, who was born in 1848. In Faulkner's imagination, however, it would have been an attractive notion to parallel the ages of these two central figures from the world of his youth.

22. "Ambuscade," p. 13; "Raid," pp. 77-78; "Skirmish at Sartoris," p. 197.

23. Jenkins (*Black-White Relations,* p. 128) formulates the problem as originating in the fact that "Ringo grows up faster than Bayard but doesn't achieve moral growth, and certainly not the fulfillment of his potential. He becomes arrested, a suitably tragic subject of treatment of no interest to Faulkner in this context."

Notes to "Clytie's Secret"

1. Compare James Kirk Paulding, *Letters from the South* (1817) and William Alexander Caruthers, *The Kentuckian in New York* (1834).

2. Harriet Martineau spoke of slavery as a "tremendous curse, the possession of irresponsible power over slaves." Quoted by William Gilmore Simms in "The Morals of Slavery," in William Gilmore Simms et al., *The Pro-Slavery Argument,* reprinted ed. (New York: Negro Universities Press, 1968), p. 215. In "The Slave Ships" (1833), John Greenleaf Whittier wrote of the crew of a slaver who threw blind slaves overboard, then were struck blind themselves by "the awful curse of God." John Thomas, ed., *Slavery Attacked: The Abolitionist Crusade* (Englewood Cliffs, N.J.: Spectrum Books, 1965), p. 40. The notion Faulkner developed of a curse that affected blacks and whites alike is closer to that voiced by Henry Clay, who called slavery a "curse to the master and a wrong to the slave." Quoted in James Ford Rhodes, *A History of the United States,* II (New York: Macmillan Co., 1928), p. 303. The moral blindness implicit in Sutpen's "design" recalls such attitudes as William Lloyd Garrison's belief that "it is morally impossible . . . for a

slaveholder to reason correctly on the subject of slavery. His mind is warped by a thousand prejudices, and a thick cloud rests upon his mental vision." "William Lloyd Garrison Abandons Colonization," in Thomas, ed., *Slavery Attacked,* p. 7. For the notion of a curse affecting several generations of a family, Faulkner had to look no further than Nathaniel Hawthorne's *The House of Seven Gables.* The biblical curse is stated in Exodus 20:5 and Deuteronomy 5:9 (both passages are identical in the King James Bible): "I the Lord thy God am a jealous God, visiting the iniquity of the fathers upon the children unto the third and fourth generation of them that hate me." For many abolitionist writers, notes George B. Tindall, "the plantation myth simply appeared in reverse, as a pattern of corrupt opulence resting upon human exploitation. Gentle old marster became the arrogant, haughty, imperious potentate, the very embodiment of sin, the central target of antislavery attack. He maintained a seraglio in the slave quarters; he bred Negroes like cattle and sold them down the river . . . , separating families . . . , while Southern women suffered in silence the guilty knowledge of their men's infidelity. The happy darkies in . . . [the] picture became white men in black skins, an oppressed people longing for freedom, the victims of countless atrocities." "Mythology: A New Frontier in Southern History," *The Idea of the South: Pursuit of a Central Theme,* ed. Frank E. Vandiver (Chicago: University of Chicago Press, 1964), p. 5.

3. Gerald Langford has argued that Faulkner never reconciled the question of which parts of Bon's story Quentin learned for himself and which he inherited from General Compson through his father. "Apparently," Langford writes, "the original idea was that the truth about Bon should be known from the beginning"; but during revision, Faulkner kept changing that. "It seems clear enough that Faulkner began with one intention, changed his mind, but then returned to his original intention. In revising the novel he changed his mind again but failed to alter several passages which indicate that the truth about Bon had been at least surmised all along." My account of Quentin's discoveries about Bon is based on the following assertion from Langford: "In spite of the inconsistencies outlined above, it is clear that Quentin is meant to become the discoverer of the fact which enables his father and himself to understand the events over which they have puzzled so long." *Faulkner's Revision of ABSALOM, ABSALOM!: A Collation of the Manuscript and the Published Book* (Austin: University of Texas Press, 1971), pp. 9, 11. Brooks, admitting "that there are a good many minor inconsistencies," argues "that the novel is essentially coherent and its parts self-consistent." The novel never shows conclusively that Bon has black ancestry, Brooks points out, but emphasizes his belief that Bon does and that Quentin learns this from his encounter with aging Henry Sutpen. *Toward Yoknapatawpha and Beyond,* pp. 302, 303-28. Shoenberg (*Old Tales and Talking,* p. 83) comments: "Langford's study indicates that in an earlier draft Mr. Compson and others *knew* Bon to be both Negro and Sutpen, but Langford does not comment on the significance of Faulkner's having changed this. That Faulkner called on Quentin to invent these 'facts' at one blow [or to discover them, one might add, as in Brooks's version, in one blow] demotes Sutpen and promotes Quentin to star billing in the cast of *Absalom, Absalom!,* for it moves the novel's arena of conflict from Sutpen's to Quentin's mind." To which should be added that this touch reinforces Quentin's preoccupation with the fears he has projected onto miscegenation and adds significantly to the climactic moment when Shreve shows him a world in which the older values are obliterated in a universal miscegenation. The most thorough

study of the significance of the information supplied by the novel's various narrators is Thomas E. Connolly, "Point of View in *Absalom, Absalom!*" *Modern Fiction Studies*, 27 (Summer 1981), 255-72. Compare also Floyd C. Watkins, "What Happens in *Absalom, Absalom!*?" *Modern Fiction Studies*, 13 (Spring 1967), 79-87; Brooks, *Yoknapatawpha Country*, pp. 436-38; Donald M. Kartiganer, *The Fragile Thread: The Meaning of Form in Faulkner's Novels* (Amherst: University of Massachusetts Press, 1979), pp. 98-102; Olga W. Vickery, *The Novels of William Faulkner*, rev. ed. (Baton Rouge: Louisiana State University Press, 1964), pp. 84-102.

4. Brown, "A Century of Negro Portraiture in American Literature," in Abraham Chapman, ed., *Black Voices* (New York: Mentor Books, 1968), p. 570. For an account of the history of the tragic mulatto stereotype, see also Penelope Bullock, "The Mulatto in American Fiction," *Phylon* (First Quarter 1945), 78-82.

5. William Sumner Jenkens, *Pro-Slavery Thought in the Old South* (Chapel Hill: University of North Carolina Press, 1935), p. 244.

6. Simms thought racial "purity" might not always be a blessing. "Perhaps the very homogeneousness of a people is adverse to the most wholesome forms of liberty," he speculated. "It may make of a selfish people . . . a *successful* people—in the merely worldly sense of the word—but it can never make them, morally, a great one." Creating superior mental and physical specimens, Simms felt, called for "strange admixtures of differing races": such was "the history of the Saxon boors under the Norman conquest—a combination, which has resulted in . . . one of the most perfect specimens of physical organization and moral susceptibilities, which the world has ever known." If the Irish had come to America and enslaved the Indians, one of Simms's characters speculated in "The Wigwam and the Cabin," they might have produced "the very noblest specimens of humanity . . . that the world had ever witnessed." But Simms could not bring himself to believe that mixing African and white strains could produce anything but degeneration, despite the fact that he himself looked on blacks as "inferior beings." "Mulatto slaves are not liked," he warned. Simms, "Morals of Slavery," in Simms et al., *Pro-Slavery*, pp. 268, 281, 238, 179.

7. *The Leopard's Spots* (New York: Doubleday, Page and Co., 1902), p. 242.

8. He apparently began "A Justice" in that year. See *FAB*, p. 566.

9. For an excellent analysis of the relationships between Bon and the Sutpen family, see Vickery, *Novels of Faulkner*, pp. 96-99; also, Ilse Dusoir Lind, "The Design and Meaning of *Absalom, Absalom!*" in *William Faulkner: Three Decades of Criticism*, ed. Frederick J. Hoffman and Olga W. Vickery (New York: Harbinger Books, 1963), pp. 281-90.

10. Brooks writes that through Shreve "Faulkner has in effect acknowledged the attitude of the modern 'liberal,' twentieth-century reader, who is basically rational, skeptical, without any special concern for history, and pretty well emancipated from the ties of family, race or section." By thus venturing to characterize liberal readers, Brooks not only disqualifies them as critics of Faulkner but also enlists Faulkner in the ranks of antiliberalism. "It was a stroke of genius on Faulkner's part," he says, "to put such a mentality squarely inside the novel, for this is a way of facing criticism from that quarter and putting it into its proper perspective." Brooks seems to have few doubts as to what that perspective is. "In fact," he writes, "Shreve sounds very much like certain literary critics who have written on Faulkner." *Yoknapatawpha Country*, p. 313. But where is the evidence that Shreve is a liberal? Certainly not where he is ranting about race: "it takes two

niggers to get rid of one Sutpen." "You've got one nigger left." "They will bleach out again like the rabbits and the birds do, so they won't show up so sharp against the snow" (*AA,* 378). Nor does Shreve, at the climactic moment when he asks Quentin why he hates the South, "resume the tone of easy banter" (Brooks, *Yoknapatawpha Country,* p. 317); it is impossible to read that passage without realizing the earnestness of Shreve's ironic probing of Quentin. Categories like liberal and conservative are not helpful with Shreve and Quentin; terms like Puritan and Cavalier offer more insight.

11. Faulkner's response to a question about *Uncle Tom's Cabin* at West Point in 1962 revealed that at some time in his career he had read that novel with care. His statement was typically ambivalent. The novel "was written out of violent and misdirected compassion and ignorance of the author toward a situation which she knew only by hearsay," he said. But he also thought Stowe was also writing "out of her heart," that "she was writing about Uncle Tom as a human being—and Legree and Eliza as human beings, not as puppets" (*FAWP,* p. 104).

12. "Quentin," concludes Shoenberg, "is a but thinly disguised autobiographical figure, and his dilemma of whether to love or to hate, to credit or to denounce, is the dilemma of his creator, the author of the book in which he exists." Faulkner's response to Quentin, she notes, is much like that of James Joyce to Dedalus or Johann von Wolfgang Goethe to young Werther. "It is worth noting that although . . . [these protagonists] differed in their reactions to the problem of impossible choice, all three did renounce and by doing so escaped their problems—simply walked away from them to Paris or to death, . . . but their authors did not. For the writers, renunciation by the proxy of their created characters was not only acceptable but preferable." *Old Tales and Talking,* p. 149. Of the roles of Quentin in *The Sound and the Fury* and *Absalom, Absalom!,* Irwin comments: "It is tempting to see in Quentin a surrogate of Faulkner, a double who is fated to retell and reenact the same story throughout his life just as Faulkner seemed fated to retell in different ways the same story again and again. . . . It is as if, in the character of Quentin, Faulkner embodied, and perhaps tried to exorcise, certain elements present in himself and in his need to be a writer." *Doubling and Incest,* p. 158.

Notes to "Letting Go"

1. *LIG,* pp. 59, 169; see also pp. 64, 191-92, 216, 234, 280, 283.

2. *FCF,* pp. 104, 111.

3. *MBB,* p. 242; *FOM,* p. 5; Cullen and Watkins, *Old Times,* p. 49.

4. Wittenberg (*Transfiguration,* p. 194) writes: "Faulkner was proud of his land acquisitions. . . . Thus his indictment of the acquisitive instinct in *Go Down, Moses* applies to himself as well as to mankind in general. . . . there is a real sense in which the book serves as both comment on and partial expiation of Faulkner's 'sins' against Caroline Barr and against the land, as well as a memorial to experiences from his past." To this one might add that the farm may have become a kind of proving ground for his intuitions about his family past, through which Faulkner tested both his pride in that past and whatever need for expiation he may have begun to feel.

5. *FAB,* p. 1243.

6. Ibid., p. 1035.

7. When Cantwell visited Rowan Oak just after *The Unvanquished* appeared in

1938, he found Faulkner thinking about his great-grandfather, and Faulkner sounded, for the moment, as though the Old Colonel himself might have invoked such a curse. Faulkner was speculating about the attenuation of the Old Colonel's feeling for his fellow humans. "He had killed two or three men," he told Cantwell. "And I suppose when you've killed men something happens inside you—something happens to your character." Did that mean that the Old Colonel might have invoked the kind of curse that had harrowed Sutpen? Like Colonel John, he was apparently trying to change. "He said he was tired of killing people," Faulkner said. "And he wasn't armed the day Thurmond shot him, although he always carried a pistol." "Recollections," p. 56.

8. The link in Faulkner's imagination between his "protagonists" is indicated by the fact that Quentin rather than Isaac is the youthful initiate in "Lion," an early version, published in 1935, of the material finalized as "The Bear." Shoenberg argues that the "characterization of Quentin [in "Lion"] suggests a still earlier writing," dating from "a time before Faulkner had begun to think of Quentin as a suicide." Shoenberg also believes that Faulkner originally had Quentin in mind in two other stories that he brought into *Go Down, Moses,* "The Old People" and "A Bear Hunt." *Old Tales and Talking,* pp. 18-19.

9. See *FAB,* pp. 1050-51. "Was" was originally titled "Almost."

10. Millgate correctly emphasizes the credence Faulkner meant to develop in Cass and the position which he develops: "Cass is not present simply as a foil to Ike; his position is no less firmly based than Ike's and in certain ways it is he who gets the better of the discussion." *Achievement of Faulkner,* p. 207.

11. Their "third generation," Faulkner wrote, wearing "hooded sheets," would lead "lynching mobs against the race their ancestors had come to save" (*GDM,* 290). He was, obviously, referring to the revived Klan, not the Nathan Bedford Forrest group of Isaac's time.

12. In "A Justice," Sam is the son of the Indian Craw-ford and a slave who is presumably a full-blooded African; in *Go Down, Moses* he is the son of Doom and a part-white slave—and thus half-Indian, but with a large portion of black and a small portion of African ancestry. For a thorough account of the backgrounds in history of Faulkner's Indian stories, see Lewis M. Dabney, *The Indians of Yoknapatawpha: A Study in Literature and History* (Baton Rouge: Louisiana State University Press, 1974).

13. Of the dialogue at the commissary, Millgate writes: "It is of the utmost importance, here as throughout Faulkner's work, not to regard a single character as a mouthpiece for the author's own views." *Achievement of Faulkner,* p. 207. It is equally important, however, to remember that many of Faulkner's views are indeed expressed by his characters, and that these are often the very beliefs which cause his fiction to take the shape it does; the question is, *which views are Faulkner's*? He apparently subscribed to some of those he ascribed to Isaac and to some that he ascribed to Cass. See ch. 12 herein.

14. *FIU,* p. 37.

15. Faulkner seemed to have forgotten Moketubbe, his symbol in "Red Leaves" of a degeneracy that might ensue when African genes were mixed with those of any other race.

16. Emphasis added.

17. Regarding the controversy over this gesture, see Brooks, *Yoknapatawpha Country,* pp. 417-20. Brooks takes issue with critics who "express their disappointment with Isaac" on this and other points: "it behooves the critic to under-

stand Isaac and not to attribute to him acts and motives which are not his." Brooks sees the gesture as an affirmative one, because "in divesting himself of his legacy [the plantation] . . . he has thereby reduced his power to act," and hence Isaac "can give only a symbolic acknowledgment—the one treasured possession that is his" (Ibid., p. 273). Arthur F. Kinney writes: "the gift is one of regeneration, passing on to the child . . . the old wilderness which he cannot know at first hand. The horn is from the good past, not the bad; it is significantly generous." "Faulkner and the Possibilities for Heroism," in *Bear, Man, and God: Eight Approaches to William Faulkner's "The Bear,"* eds. Francis Lee Utley, Lynn Z. Bloom, and Kinney (New York: Random House, 1971), p. 241.

18. Dixon's analogy was that the character of black people, like their pigmentation, was as unalterable as the spots of the leopard—hence the title of his book, *The Leopard's Spots.*

19. Faulkner told Malcolm Cowley that "Rider was one of the McCaslin Negroes" (*FCF,* p. 113). Millgate argues that "Faulkner's refusal to make the few minor changes in names and relationships which would have made Rider a McCaslin has the effect not of isolating the episode in which Rider is the major character but actually of expanding, beyond the limits of the McCaslin family, the whole scope and relevance of the book." *Achievement of Faulkner,* p. 204. This is important; but to have made Rider "one of the McCaslin Negroes" is not the same thing as to make "changes . . . which would have made Rider a McCaslin." Making him a McCaslin would have been to imply at least that Rider was part-white; by declining to make him one, Faulkner allowed himself to suggest that Rider had no significant portion of white ancestry, thus isolating both his character and his tragedy as types Faulkner could suggest as typical of "pure" black Americans.

20. Faulkner's original concept of Lucas was very close to the stereotype of the comic darky. For an account of the development of his attitudes toward Lucas, see James Early, *The Making of* Go Down, Moses (Dallas: Southern Methodist University Press, 1972), pp. 3-10.

21. In *Go Down, Moses,* as in Faulkner's other works, blacks who have white ancestry are usually identified by their known connections to white ancestors; blacks without significant non-African ancestry are identified only by reference to pigmentation.

22. *MBB,* p. 48.

Notes to "A Once and Future South"

1. To Robert K. Haas, *SLWF,* p. 122.
2. To Harold Ober, ibid., p. 199.
3. To Malcolm A. Franklin, ibid., p. 166.
4. To Robert K. Haas, ibid., p. 180.
5. Shortly after *Intruder in the Dust* appeared, Edmund Wilson reviewed it for the *New Yorker* (24 [Oct. 23, 1948], 120-22, 125-28) as "Faulkner's Reply to the Civil Rights Program." In 1963 Brooks, noting a "kind of reader" who "comes to the novel with certain expectations and certain hostilities and irritabilities," observed: "In the charged atmosphere of today (and of 1948, when the novel appeared) such bias is to be expected. It becomes necessary, then, to remind the reader that *Intruder in the Dust* is a novel and not, in spite of Edmund Wilson's comments, a tract." *Yoknapatawpha Country,* pp. 280-81. However, Millgate,

admitting that "it seems unlikely" that "Faulkner's primary intention" was "a work of propaganda," writes that "the whole pattern of the novel appears to be manipulated in such a way as to confirm and ratify the validity of [Gavin] Stevens's concluding statements, and . . . [Charles Mallison's] actions themselves become the practical text for his uncle's sermons on the South." *Achievement of Faulkner,* pp. 215-16. Charles D. Peavey writes that "*Intruder in the Dust* is both novel and tract." *Go Slow Now: Faulkner and the Race Question* (Eugene: University of Oregon Press, 1971), p. 46.

6. To Harold Ober, *SLWF,* p. 262.

7. Quoted in *FAB,* pp. 1245-46.

8. *FIU,* p. 142. For a detailed account of the composition of *Intruder in the Dust,* see Patrick H. Samway, *Faulkner's* Intruder in the Dust: *A Critical Study of the Typescripts* (Troy, N.Y.: Whitston Publishing Co., 1980), pp. 1-33.

9. In *Sartoris* Frenchman's Bend was hill country and in *The Hamlet* (p. 5) Faulkner observed that "there was not one Negro landowner in the entire [Frenchman's Bend] section. Strange Negroes would absolutely refuse to pass through it after dark."

10. *FCF,* p. 111.

11. Faulkner was somewhat inconsistent. In *Intruder in the Dust* Ephraim thought "young folks and womens aint cluttered"; but Faulkner told Cowley, "It's the grown-ups and especially the women who keep the prejudice alive." *FCF,* p. 111.

12. In *Go Down, Moses* Molly's white childhood companion was named Worsham.

13. *LIG,* pp. 58, 56, 54.

14. To Secretary, American Academy, *FCF,* p. 141. Bleikasten argues that the "failure" of which Faulkner spoke on several occasions was an artistic one. Of Faulkner's feelings about his "failure" in *The Sound and the Fury,* Bleikasten writes: "There had been others before; with this book, however, Faulkner met failure in a deeper, more inescapable sense—failure as the very destiny of all artistic endeavor. What then became evident to him was the sobering truth that, as Samuel Beckett put it, 'to be an artist is to fail, as no other dare fail' and that 'failure is his world and the shrink from it desertion.' " Assigning *The Sound and the Fury* to "a major place in what has been, since Hawthorne, Poe, and Melville, the great tradition of failure in American literature," Bleikasten asserts that in the careers of such writers, "everything happens as though the writing process could never be completed, as though it could only be the gauging of a lack. Creation then ceases to be a triumphant gesture of assertion; it resigns itself to be the record of its errors, trials and defeats, the chronicle of its successive miscarriages." *Splendid Failure,* pp. 48, 50. To reduce the problem of failure to the problem of failure to achieve *artistic* success robs the artist of his motives in writing and prevents an analysis of the reasons for artistic failure. In other respects, this is a fair statement of the dynamics of the composition of *The Sound and the Fury,* and it also speaks to the dynamics of the development of Faulkner's career.

15. *FCF,* pp. 159, 110. Of his "voices," Faulkner told Cowley, "Sometimes I don't like what they say . . . but I don't change it." To Cowley, that meant that "their message might be in conflict with his [Faulkner's] conscious standards"—but that, apparently for artistic reasons, he had characters saying what their characterization called for them to say.

16. There has been much discussion as to whether Gavin Stevens is a spokes-

man for Faulkner's own beliefs. Brooks (*Yoknapatawpha Country*, p. 424), noting that Faulkner has taken positions publicly which are similar to some of Gavin's and "at variance" with others, insists that "Gavin Stevens is not Faulkner's mouthpiece. . . . he is treated by Faulkner with great detachment. His theories and arguments are not privileged utterances, but have to take their chances in the total artistic context." However, any subtle propagandist separates himself from his creations by allowing them to enunciate minor points on which they differ with their creator while they are allowed to be persuasive on more important ones on which creator and creations agree. By the time Brooks was writing (1963), Faulkner had been on record a number of times with statements about race so remarkably similar to Gavin's as to leave little doubt that he for a time at least had either seriously pondered or adopted most of them. See ch. 12 herein. Millgate (*Achievement of Faulkner*, pp. 215, 218) puts the point more subtly. "It is true that Stevens is a created character with whose views Faulkner cannot be directly identified; it is also true that in the early stages of the novel Stevens suffers from . . . limitations of his time, class, and environment. . . . But Stevens takes to himself the truths his nephew discovers, absorbs them into his thinking, and speaks in the later sections of the novel with the authority of this new wisdom. . . . The whole pattern of the novel appears to be manipulated in such a way as to confirm and ratify the validity of Stevens's concluding statements." For Millgate, the novel is more than propaganda. By forcing readers to see the action through the eyes of Chick, Faulkner "forces . . . [them] to view the events of the novel in terms of his own experience of them, an experience which we are apparently intended to see as a progressive initiation into manhood and for which the predicament of Lucas Beauchamp provides not so much the cause as the context." But if a novel is somehow more than propaganda, what is to be made of the propaganda that it contains? Faulkner's method, as his comments on characters such as Quentin, Bayard Sartoris of *The Unvanquished*, Isaac McCaslin, Chick, and Gavin reveal, is to create characters whose visions of the South represent his own fears and hopes, then to test the vision. A Quentin or an Isaac or a Gavin represents Faulkner asking a question about the limits of his own vision at a particular juncture, which he proceeds to probe through his plot. When the test is over, he proceeds to his next figure, whose vision may be larger than that of his predecessor. Gavin voices Faulkner's deep concern at a climactic moment in his career; the depth of his concern is evident in that Faulkner allowed no opposing voice to express itself freely in *Intruder in the Dust* on most important issues and that Faulkner did later express as his own many of the beliefs enunciated by Gavin.

17. Brooks (*Yoknapatawpha Country*, p. 421) writes that Gavin means cultural, not racial, homogeneity: "Gavin cannot mean by the term a sameness in racial stock though this is what the term will suggest to most readers. Gavin is aware that the white Southerners represent a blending of somewhat different racial stocks, as his own references in this book make clear. And the 'confederation' of two 'homogeneous' peoples, the Negroes and the white Southerners, for which he asks is certainly not the uniting of two branches of the same racial stock. In calling a people homogeneous Gavin can only mean that they have a community of values that is rooted in some kind of lived experience." Such a view presents the following problems. (1) To say that "homogeneity" suggests something different to most readers from what Faulkner intended is to deny Faulkner either his intentions or his knowledge of what the word means. (2) To say that

Gavin does not mean one thing because elsewhere he means another is merely to argue consistency from inconsistency. (3) Who could possibly think Faulkner meant southern blacks and whites were "two branches of the same racial stock"? This diverts attention from the issue. (4) Since genetic racial "homogeneity" has never existed, Brooks's statement that Gavin is aware of "a blending of somewhat different racial stocks" is misleading; such "a blending" is all that "homogeneity" could ever be in the first place. If the word "ethnic" is used instead of "genetic" and "racial," both Gavin's and Brooks's statements become clearer. Ethnic "homogeneity" would not be dependent on race (though it is frequently taken to be); it is ethnic homogeneity that Gavin believes will alone produce "anything of a people . . . of durable and lasting value" (*ID*, 154).

18. Brooks (*Yoknapatawpha Country*, p. 280) summarizes the problems: (1) "the extravagances of the plot," which in "a good detective story" has to "justify its complications," and (2) "the incoherence of motivation and action"—particularly as regards the murder of Vinson Gowrie and the framing of Lucas.

Notes to "Doing Something about It"

1. *ESPL*, p. 124.
2. Ibid., p. 120.
3. *FIU*, pp. 245-46. See also *LIG*, p. 247. Of this development in Faulkner's career, Wittenberg finds an "expansionary quality" manifested in *The Sound and the Fury* which is "a paradigm for Faulkner's career as a novelist," explaining that "as he moved increasingly outward from private to public issues [in *The Sound and the Fury*], the work lessened in its psychological intensity." This development eventuated in situations in *Go Down, Moses* and *The Hamlet* in which he "began to move away from his personal obsessions and to concentrate instead on his various responses to social issues." *Transfiguration*, pp. 84, 191. One could say that Faulkner's preoccupation with personal obsessions grew originally from his perception of his own relationship and that of his family to social issues and that the movement in his fiction was from the personal to the social. Viewed from this perspective, Faulkner's efforts to forestall what he viewed as a social crisis over race represent a continuum of a movement begun in his fiction. If this is true, they demand the same degree of attention as his works of art, representing as they do an outgrowth of Faulkner's psychic development begun in his works.
4. *ESPL*, pp. 146-48, 151.
5. *FIU*, pp. 25-26.
6. *ESPL*, p. 82.
7. *LIG*, p. 167.
8. Ibid., p. 101. Members of the Falkner family left different—and sometimes conflicting—accounts of their beliefs on the subject of ancestors, but Colonel W. C. Falkner left no record of his own. Sallie Murry Falkner, wife of J. W. T. Falkner, brought both Murry and McAlpine blood to the Falkners, according to family tradition. "Thus," concludes Blotner (*FAB*, p. 5), "the Scottish antecedents begin to appear clearly in the generation following that of the [Old] Colonel." See ibid., pp. 3-7.
9. *LIG*, p. 191.
10. Ibid., pp. 159-60. Faulkner's original statement contains a double negative, which makes it unintentionally ambiguous: "I think that one never loves a land, a people, or even a person not because of what that person or land or place is but despite what that person or place is."

11. Faulkner later repudiated parts of this interview; see the subsequent discussion in this chapter.

12. Ibid., pp. 264, 263.

13. Ibid., pp. 183, 264, 263.

14. *ESPL*, p. 149. Faulkner seemed pleased with this passage, using parts of it again in "On Fear: The South in Labor" (first published in *Harper's*, 212 [June 1956], 30). The essay was reprinted as "On Fear: Deep South in Labor: Mississippi, 1956," *ESPL*, pp. 92-106.

15. *ESPL*, p. 87.

16. *LIG*, p. 90.

17. *ESPL*, p. 109.

18. *LIG*, pp. 183, 184, 182.

19. Ibid., p. 143.

20. Ibid., pp. 184, 89, 161, 183.

21. *FAB*, p. 1314. See also Peavey, *Go Slow Now*, pp. 50-53.

22. *ESPL*, pp. 203-4.

23. Ibid., pp. 221, 219, 216, 220-21. See also Peavey, *Go Slow Now*, pp. 58-65.

24. *ESPL*, pp. 222-23. See also Peavey, *Go Slow Now*, pp. 65-68.

25. *ESPL*, pp. 86, 87. See also Peavey, *Go Slow Now*, pp. 75-78.

26. *ESPL*, pp. 86-87, 90.

27. Ibid., pp. 88, 89, 91.

28. Quoted in "On Fear," ibid., p. 93.

29. Quoted in Bradford A. Daniel, "William Faulkner and the Southern Quest for Freedom," in his *Black, White and Gray: Twenty-one Points of View on the Race Question* (New York: Sheed and Ward, 1964), pp. 295-96.

30. *FIU*, p. 53.

31. *MBB*, p. 268.

32. *FOM*, p. 199.

33. *MBB*, p. 268. See also Peavey, *Go Slow Now*, pp. 74-75; *FAB*, p. 1532.

34. Hodding Carter, "Faulkner and His Folk," *Princeton University Library Chronicle*, 18 (Spring 1957), 106.

35. *Life*, 40 (Mar. 26, 1956), 19; (Apr. 16, 1956), 21.

36. New York *Times*, Apr. 18, 1956. See also Peavey, *Go Slow Now*, pp. 77-78.

37. James Baldwin, "Faulkner and Desegregation," *Partisan Review*, 23 (Winter 1956), 570.

38. *ESPL*, p. 101.

39. *FAB*, pp. 1590, 1592.

40. Ibid., p. 1590. See also Peavey, *Go Slow Now*, pp. 68-74. According to Blotner, "When he gave . . . [the interview with Howe] he had been drinking steadily but not enough to prevent his making comments more desperate and provocative than any he had thus far uttered."

41. *LIG*, pp. 262, 263, 259, 260.

42. Ibid., pp. 259-62.

43. Miss Maud's stroke occurred on Oct. 23, 1955 (*FAB*, p. 1580). Mrs. Lida Oldham, Estelle's mother, died Mar. 10, 1956. The fall and the incident with the youth, Charles Moorer, took place in early or mid-March (*FAB*, pp. 1597-99). Blotner (ibid., p. 1598) concludes that Faulkner's hospitalization was the result of "accumulated anger, fear, frustration, injury, sickness, and drinking."

44. *Time*, 67 (Apr. 23, 1956), 2. He wrote similarly worded statements that appeared in *The Reporter*, 14 (Apr. 19, 1956), 7, and "If I Were a Negro," *ESPL*,

p. 107. See also Peavey, *Go Slow Now,* pp. 71-72.

45. *The Reporter,* 14 (Apr. 19, 1956), 7. Years later, Faulkner's friend, Jean Stein, would argue that "given the circumstances of the review and Mr. Faulkner's subsequent statement, I believe that History should accept the latter" (*FAB,* p. 1600).

46. *ESPL,* pp. 108, 109, 110.

47. Ibid., p. 149.

48. Ibid., p. 112.

49. *FIU,* pp. 222, 221, 217.

50. Ibid., p. 213. The complete statement reads: "I think that all that don't violently repudiate what I've said will agree that the white man is responsible for the Negro's condition, the fact that the Negro does act like a Negro and can live among us and be irresponsible."

51. Ibid., pp. 210-11. Faulkner qualified these statements by asking hypothetical northerners to agree with them "for the sake of the argument"; but nine months earlier he had felt no need for such a qualification when he stated: "If . . . [equality] is given to [the black] by a Supreme Court ukase and enforced with police, as soon as the police are gone then some smart white man, maybe some smart Negro, will take his equality away from him again. He has got to be taught the responsibilities of equality" (ibid., p. 148).

52. Ibid., p. 210.

53. Sutpen came from the mountain country of what is now West Virginia, although as a youth he had walked down to the Tidewater with his family. An early version of *Sartoris* had a Sartoris who was a contemporary of Charles I coming to Virginia, but by the time the manuscript was complete that portion had been cut.

54. Ibid., pp. 221-22, 212.

55. Ibid., p. 282.

56. *LIG,* pp. 179, 180. Compare ibid., pp. 88-89, 121-22, 225, 238; *FIU,* pp. 143-44, 206.

57. *FIU,* pp. 98, 68.

Notes to "Home"

1. *FAB,* pp. 1787-88.

2. Ibid., p. 1787.

3. The episode was taken from a short story that Faulkner had completed in 1954, "By the People," which first appeared in *Mademoiselle.*

4. *FAB,* p. 1795. The friend was Joseph Blotner, who was allowed to read some of the manuscript during composition.

5. For an account of the family's adventures with the Buick, see *FOM,* pp. 62-77. Included in these pages is a photograph of J. W. T. Falkner standing beside the Buick, which is stuck in a mud hole.

6. Though Boon was one of the significant participants in the hunts in "The Bear," his influence on Isaac McCaslin was indirect, never dramatized, and Faulkner seldom portrayed the two as spiritually or emotionally close. The most significant occasions when the two were in contact were a trip to Memphis, in which Boon ignored Isaac and occupied himself drinking, and the death of Sam Fathers, in which Boon and Isaac gave Sam an Indian burial.

7. Millgate speaks for a number of critics when he identifies the novel's "treatment of white-Negro relations" as "varied but unstrenuous," arguing that *The*

Reivers should not be read as a statement about relations between the races. "We are never allowed to forget that the action is not contemporary, and the whole strategy of the narrative point of view serves to stress the character of the novel as an affectionate evocation of a world that cannot return. Seen from this angle, it becomes irrelevant to speak of the portrait of Uncle Parsham as sentimentalized, or to complain of the obsolescence of the standards by which white behavior towards Negroes appears to be judged." *Achievement of Faulkner,* p. 256. This view ignores one of the major thrusts of proslavery propaganda, which, since John Pendleton Kennedy's *Swallow Barn* (1832), has sought to obscure the horrors of the peculiar institution behind a cloud of nostalgia. William R. Taylor has traced the history of this species of reminiscence in *Cavalier and Yankee: The Old South and National Character* (New York: George Braziller, 1961).

Selected Bibliography

BOOKS

Adams, Richard P. *Faulkner: Myth and Motion.* Princeton, N.J.: Princeton University Press, 1968.

Anderson, Sherwood. *The Letters of Sherwood Anderson,* ed. Howard Mumford Jones and Walter B. Rideout. Boston: Little, Brown and Co., 1953.

Bachman, Melvin. *Faulkner: The Major Years.* Bloomington: Indiana University Press, 1966.

Baldwin, James. *Notes of a Native Son.* Boston: Beacon, 1955.

Bassett, John, ed. *William Faulkner: The Critical Heritage.* London: Routledge & Kegan Paul, 1975.

Beck, Warren. *Man in Motion: Faulkner's Trilogy.* Madison: University of Wisconsin Press, 1961.

Bleikasten, André. *The Most Splendid Failure.* Bloomington: Indiana University Press, 1976.

Blotner, Joseph. *Faulkner: A Biography.* Vols. I and II. New York: Random House, 1974.

———, ed. *Selected Letters of William Faulkner.* New York: Random House, 1977.

Broughton, Panthea Reid. *William Faulkner: The Abstract and the Actual.* Baton Rouge: Louisiana State University Press, 1974.

Brooks, Cleanth. *William Faulkner: The Yoknapatawpha Country.* New Haven, Conn.: Yale University Press, 1963.

———. *William Faulkner: Toward Yoknapatawpha ând Beyond.* New Haven, Conn.: Yale University Press, 1978.

Brylowski, Walter. *Faulkner's Olympian Laugh: Myth in the Novels.* Detroit: Wayne State University Press, 1968.

Cash, Wilbur J. *The Mind of the South.* New York: Alfred A. Knopf, 1941.

Cleaver, Eldridge. *Soul on Ice.* New York: Delta Books, 1968.

Cowley, Malcolm, ed. *The Faulkner-Cowley File: Letters and Memories, 1944-1962.* New York: Viking, 1966.

Cullen, John B., and Floyd C. Watkins. *Old Times in the Faulkner Country.* Chapel Hill: University of North Carolina Press, 1961.

Dabney, Lewis M. *The Indians of Yoknapatawpha: A Study in Literature and History.* Baton Rouge: Louisiana State University Press, 1974.

Dixon, Thomas, Jr. *The Clansman.* New York: Doubleday, Page and Co., 1905.
———. *The Leopard's Spots.* New York: Doubleday, Page and Co., 1902.
Du Bois, William Edward Burghardt. *The Souls of Black Folk: Essays and Sketches.* New York: Signet Classics, 1969.
Early, James. *The Making of* Go Down, Moses. Dallas: Southern Methodist University Press, 1972.
Ellison, Ralph. *Invisible Man.* New York: Random House, 1952.
———. *Shadow and Act.* New York: Random House, 1964.
Fadiman, Regina K. *Faulkner's* Light in August: *A Description and Interpretation of the Revisions.* Charlottesville: University Press of Virginia, 1975.
Falkner, Murry C. *The Falkners of Mississippi.* Baton Rouge: Louisiana State University Press, 1967.
Falkner, William Clark. *The White Rose of Memphis.* New York: Bond Wheelwright Co., 1953.
Fant, Joseph L., and Robert Ashley, eds. *Faulkner at West Point.* New York: Random House, 1964.
Faulkner, John. *My Brother Bill: An Affectionate Reminiscence.* New York: Trident Press, 1963.
Faulkner, William. *Essays, Speeches and Public Letters.* Ed. James B. Meriwether. New York: Random House, 1965.
Fiedler, Leslie. *Love and Death in the American Novel.* New York: Criterion Books, 1960.
Gwynn, Frederick L., and Joseph Blotner, eds. *Faulkner in the University.* Charlottesville: University of Virginia Press, 1959.
Holman, Clarence Hugh. *Three Modes of Southern Fiction: Ellen Glasgow, William Faulkner, Thomas Wolfe.* Athens: University of Georgia Press, 1966.
Holmes, William F. *The White Chief: James Kimble Vardaman.* Baton Rouge: Louisiana State University Press, 1970.
Howe, Irving. *William Faulkner: A Critical Study.* 2nd ed. New York: Vintage, 1952.
Huggins, Nathan Irvin. *Harlem Renaissance.* New York: Oxford University Press, 1971.
Irwin, John T. *Doubling and Incest/Repetition and Revenge: A Speculative Reading of Faulkner.* Baltimore, Md.: Johns Hopkins University Press, 1975.
Jehlen, Myra. *Class and Character in Faulkner's South.* New York: Columbia University Press, 1976.
Jenkens, William Sumner. *Pro-Slavery Thought in the Old South.* Chapel Hill: University of North Carolina Press, 1935.
Jenkins, Lee. *Faulkner and Black-White Relations: A Psychoanalytic Approach.* New York: Columbia University Press, 1981.
Johnson, James Weldon. *The Autobiography of an Ex-Colored Man.* New York: Alfred A. Knopf, 1970.
Kartiganer, Donald M. *The Fragile Thread: The Meaning of Form in Faulkner's Novels.* Amherst: University of Massachusetts Press, 1979.
Kerr, Elizabeth M. *William Faulkner's Gothic Domain.* Port Washington, N.Y.: Kennikat Press, 1979.
———. *Yoknapatawpha: Faulkner's "Little Postage Stamp of Native Soil."* 2nd ed. rev. New York: Fordham University Press, 1976.

Kinney, Arthur F. *Faulkner's Narrative Poetics: Style as Vision.* Amherst: University of Massachusetts Press, 1978.

Kirwan, Albert D. *Revolt of the Rednecks.* Lexington: University of Kentucky Press, 1951.

Langford, Gerald. *Faulkner's Revision of ABSALOM, ABSALOM!: A Collation of the Manuscript and the Published Book.* Austin: University of Texas Press, 1971.

Levins, Lynn Gartrell. *Faulkner's Heroic Design: The Yoknapatawpha Novels.* Athens: University of Georgia Press, 1976.

Meriwether, James B. *The Literary Career of William Faulkner.* Princeton, N.J.: Princeton University Press, 1961.

———, and Michael Millgate, eds. *Lion in the Garden: Interviews with William Faulkner, 1926-1962.* New York: Random House, 1968.

Millgate, Michael. *The Achievement of William Faulkner.* New York: Random House, 1966.

Miner, Ward L. *The World of William Faulkner.* Durham, N.C.: Duke University Press, 1952.

Minter, David. *William Faulkner: His Life and Work.* Baltimore, Md.: Johns Hopkins University Press, 1980.

Murphy, James B. *L. Q. C. Lamar: Pragmatic Patriot.* Baton Rouge: Louisiana State University Press, 1973.

Peavey, Charles D. *Go Slow Now: Faulkner and the Race Question.* Eugene: University of Oregon Press, 1971.

Percy, William Alexander. *Lanterns on the Levee.* New York: Alfred A. Knopf, 1941.

Samway, Patrick H. *Faulkner's* Intruder in the Dust: *A Critical Study of the Typescripts.* Troy, N.Y.: Whitston Publishing Co., 1980.

Shoenberg, Estella. *Old Tales and Talking.* Binghamton, N.Y.: University Press of Mississippi, 1977.

Simms, William Gilmore et al. *The Pro-Slavery Argument.* Reprinted ed. New York: Negro Universities Press, 1968.

Slatoff, Walter J. *Quest for Failure: A Study of William Faulkner.* Ithaca, N.Y.: Cornell University Press, 1960.

Stonum, Gary Lee. *Faulkner's Career: An Internal Literary History.* Ithaca, N.Y.: Cornell University Press, 1979.

Swiggart, Peter. *The Art of Faulkner's Novels.* Austin: University of Texas Press, 1962.

Taylor, William R. *Cavalier and Yankee: The Old South and National Character.* New York: George Braziller, 1961.

Thompson, Lawrance. *William Faulkner: An Introduction and Interpretation.* New York: Barnes and Noble, 1967.

Tischler, Nancy M. *Black Masks: Negro Characters in Modern Southern Fiction.* University Park: University of Pennsylvania Press, 1969.

Vickery, Olga W. *The Novels of William Faulkner.* Rev. ed. Baton Rouge: Louisiana State University Press, 1964.

Waggoner, Hyatt. *William Faulkner: From Jefferson to the World.* Lexington: University of Kentucky Press, 1966.

Watkins, Floyd C. *The Flesh and the Word: Eliot, Hemingway, Faulkner.* Nashville, Tenn.: Vanderbilt University Press, 1971.

Watson, James Gray. *The Snopes Dilemma: Faulkner's Trilogy.* Coral Gables, Fla.: University of Miami Press, 1968.

Webb, James W., and A. Wigfall Green. *William Faulkner of Oxford.* Baton Rouge: Louisiana State University Press, 1965.

Wharton, Vernon L. *The Negro in Mississippi.* New York: Harper and Row, 1965.

Williams, David. *Faulkner's Women: The Myth and the Muse.* Montreal: McGill-Queens University Press, 1977.

Wittenberg, Judith Bryant. *Faulkner: The Transfiguration of Biography.* Lincoln: University of Nebraska Press, 1979.

Woodward, C. Vann. *The Strange Career of Jim Crow.* 2nd ed. New York: Oxford University Press, 1966.

Wright, Richard. *Black Boy: A Record of Childhood and Youth.* New York: Harper and Brothers, 1945.

———. *Native Son.* New York: Modern Library, 1940.

ARTICLES AND ESSAYS

Adams, Richard P. "The Apprenticeship of William Faulkner." *Tulane Studies in English,* 12 (1962), 124.

Anderson, Sherwood. "A Meeting South." *Sherwood Anderson Reader.* Boston: Houghton Mifflin, 1947. Pp. 274-84.

Bachman, Melvin. "Sutpen and the South: A Study of *Absalom, Absalom!*" *PMLA,* 80 (Dec. 1965), 596-604.

Baldwin, James. "Faulkner and Desegregation." *Partisan Review,* 23 (Fall 1956), 568-73.

Blanchard, Margaret. "The Rhetoric of Communion: Voice in *The Sound and the Fury.*" *American Literature,* 41 (Jan. 1970), 555-65.

Blotner, Joseph. "The Falkners and the Fictional Families." *Georgia Review,* 30 (Fall 1976), 572-92.

Bradford, M. E. "Faulkner, James Baldwin, and the South." *Georgia Review,* 20 (Winter 1966), 431-41.

Brooks, Cleanth. "The Narrative Structure of *Absalom, Absalom!*" *Georgia Review,* 29 (Summer 1975), 366-94.

Brown, Sterling A. "A Century of Negro Portraiture in American Literature." In *Black Voices.* Ed. Abraham Chapman. New York: Mentor Books, 1968. Pp. 564-89.

———. "Negro Character as Seen by White Authors." In *Dark Symphony: Negro Literature in America.* Ed. James A. Emanuel and Theodore L. Gross. New York: The Free Press, 1968. Pp. 564-89.

Bullock, Penelope. "The Mulatto in American Fiction." *Phylon,* 11 (First Quarter 1945), 78-82.

Cantwell, Robert. "The Falkners: Recollections of a Gifted Family." In *William Faulkner: Three Decades of Criticism.* Ed. Frederick J. Hoffman and Olga W. Vickery. New York: Harbinger Books, 1963. Pp. 51-66.

———. "Introduction." In William Clark Falkner, *The White Rose of Memphis.* New York: Bond Wheelwright Co., 1953. Pp. v-xxvii.

Carter, Hodding. "Faulkner and His Folk." *Princeton University Library Chronicle,* 18 (Spring 1957), 106.

Connolly, Thomas E. "Point of View in *Absalom, Absalom!*" *Modern Fiction Studies,* 27 (Summer 1981), 255-72.

Cowley, Malcolm. "The Etiology of Faulkner's Art." *Southern Review,* 13 (Winter 1977), 83-95.

Dahl, James. "A Faulkner Reminiscence: Conversations with Mrs. Maud Falkner." *Journal of Modern Literature,* 3 (1974), 1028.

Daniel, Bradford A. "William Faulkner and the Southern Quest for Freedom." In *Black, White and Gray: Twenty-one Points of View on the Race Question.* New York: Sheed and Ward, 1964. Pp. 291-308.

Hagopian, John V. "*Absalom, Absalom!* and the Negro Question." *Modern Fiction Studies,* 19 (Summer 1973), 207-11.

Hartman, Geoffrey. "Towards Literary History." In *Beyond Formalism.* New Haven, Conn.: Yale University Press, 1970. Pp. 356-86.

Inge, Thomas. "The Virginia Face of Faulkner." *Virginia Cavalcade,* 24 (Summer 1974), 32-39.

Kent, George E. "The Black Woman in Faulkner's Works with the Exclusion of Dilsey." Parts I and II. *Phylon,* 35 (Dec. 1974), 430-41; 36 (Mar. 1975), 55-67.

Key, V. O., Jr. "Mississippi: The Delta and the Hills." In his *Southern Politics.* New York: Alfred A. Knopf, 1949. Pp. 229-53.

Kinney, Arthur F. "Faulkner and the Possibilities for Heroism." *The Southern Review,* 6 (Autumn 1970), 1110-25.

Lewis, R. W. B. "The Hero in the New World: William Faulkner's *The Bear.*" *Kenyon Review,* 13 (Autumn 1951), 641-60.

Lind, Ilse DuSoir. "The Design and Meaning of *Absalom, Absalom!*" *PMLA,* 70 (Dec. 1955), 887-912.

Lovelace, Rev. C. C. "The Wounds of Jesus." In Zora Neale Hurston, *Jonah's Gourd Vine.* Philadelphia: Lippincott, 1971. Pp. 269-81.

McHaney, Thomas L. "Anderson, Hemingway, and Faulkner's *The Wild Palms.*" *PMLA,* 87 (May 1972), 465-74.

———. "The Elmer Papers: Faulkner's Comic Portraits of the Artist." *Mississippi Quarterly,* 26 (Summer 1973), 281-311.

———. "The Falkners and the Origin of Yoknapatawpha County: Some Corrections." *Mississippi Quarterly,* 25 (Summer 1972), 249-64.

Millgate, Michael. "Faulkner and the Literature of the First World War." *Mississippi Quarterly,* 26 (Summer 1973), 387-93.

———. "Starting Out in the Twenties: Reflections on *Soldiers' Pay.*" *Mosaic,* 7 (Fall 1973), 1-14.

Milum, Richard A. "Faulkner and the Cavalier Tradition: The French Bequest." *American Literature,* 45 (Jan. 1974), 580-89.

O'Donnell, George Marion. "Faulkner's Mythology." In *Faulkner.* Ed. Robert Penn Warren. Englewood Cliffs, N.J.: Prentice-Hall, 1966. Pp. 23-33.

Peavey, Charles D. "If I'd Just Had a Mother." *Literature and Psychology,* 23 (1973), 114-21.

Petesch, D. A. "Faulkner Negroes: The Conflict Between [*sic*] the Republic Man and Remote Art. *Southern Humanities Review,* 10 (Winter 1976), 55-63.

Polk, Noel. "Review Essay of the Manuscript of *Absalom, Absalom!*" *Mississippi Quarterly,* 25 (1972), 358-67.

Reed, W. McNeill. "Four Decades of Friendship." In *William Faulkner of Oxford.* Ed. James W. Webb and A. Wigfall Green. Baton Rouge: Louisiana State University Press, 1965. Pp. 180-88.

Rosenberg, Bruce A. "The Oral Quality of Reverend Shegog's Sermon in William Faulkner's *The Sound and the Fury.*" *Literature in Wissenschaft und Unterricht,* 2 (1969), 73-88.

Simpson, Lewis P. "Ike McCaslin and the Second Fall of Man." In *Bear, Man and God: Eight Approaches to William Faulkner's "The Bear."* Ed. Francis Lee Utley, Lynn Z. Bloom, and Arthur F. Kinney. 2nd ed. New York: Random House, 1971. Pp. 202-9.

Stewart, David H. "Ike McCaslin, Cop-Out." *Criticism*, 3 (Fall 1961), 333-42.

Stone, Phil. "The Man and the Land." In James W. Webb and A. Wigfall Green, *William Faulkner of Oxford.* Baton Rouge: Louisiana State University Press, 1965. Pp. 3-9.

Taylor, Walter. "Faulkner: Nineteenth-Century Notions of Racial Mixture and the Twentieth-Century Imagination." *South Carolina Review*, 10 (Nov. 1977), 57-68.

———. "Faulkner: Social Commitment and the Artistic Temperament." *Southern Review*, 6 (Autumn 1970), 1075-92.

———. "Faulkner's Curse." *Arizona Quarterly*, 28 (Winter 1972), 333-38.

———. "Faulkner's Pantaloon: The Negro Anomaly at the Heart of *Go Down, Moses.*" *American Literature*, 44 (Nov. 1972), 430-44.

———. "The Freedman in *Go Down, Moses*: Historical Fact and Imaginative Failure." *Ball University Forum*, 8 (Winter 1967), 3-7.

———. "Horror and Nostalgia: The Double Perspective of Faulkner's 'Was.' " *Southern Humanities Review*, 8 (Winter 1974), 74-84.

———"Let My People Go: The White Man's Heritage in *Go Down, Moses.*" *South Atlantic Quarterly*, 58 (Winter 1959), 20-32.

Tindall, George B. "Mythology: A New Frontier in Southern History." In *The Idea of the South: Pursuit of a Central Theme.* Ed. Frank E. Vandiver. Chicago: University of Chicago Press, 1964. Pp. 1-15.

Tischler, Nancy M. "William Faulkner and the Southern Negro." *Susquehanna University Studies*, 7 (June 1965), 261-65.

Turner, Arlin. "William Faulkner, Southern Novelist." *Mississippi Quarterly*, 14 (Summer 1961), 117-30.

Utley, Francis Lee. "Pride and Humility: The Cultural Roots of Isaac McCaslin." In *Bear, Man and God: Eight Approaches to William Faulkner's "The Bear."* Ed. Francis Lee Utley, Lynn Z. Bloom, and Arthur F. Kinney. 2nd ed. New York: Random House, 1971. Pp. 167-87.

Warren, Robert Penn. "Faulkner: The South and the Negro." *The Southern Review*, 1 (Summer 1965), 501-29.

Watkins, Floyd C. "What Happens in *Absalom, Absalom!*?" *Modern Fiction Studies*, 13 (Spring 1967), 79-87.

Wilson, Edmund. "Faulkner's Reply to the Civil Rights Program." *New Yorker*, 24 (Oct. 23, 1948), 120-22, 125-28.

Wright, Richard. "How Bigger Was Born." In *Native Son.* New York: Perennial Classics, 1966. Pp. xvii-xxxiv.

DISSERTATIONS AND THESES

Duclos, Donald P. "Son of Sorrow: The Life, Works and Influence of Colonel William C. Falkner, 1825-1889." Ph.D. diss. University of Michigan, 1962.

Wells, Dean Faulkner. "Dean Swift Faulkner: A Biographical Study." Master's thesis. University of Mississippi, 1975.

Index

Note on the Author

Walter Taylor is professor of English at the University of Texas at El Paso. A native of Clinton, Mississippi, he received his B.A. from the University of Mississippi in 1951 and his Ph.D. from Emory University in 1964. He has published scholarly articles on William Faulkner in journals such as *South Atlantic Quarterly, Southern Review, American Literature,* and *Southern Humanities Review.*